Canadian Resource Supplement

Carpentry

Fifth Edition

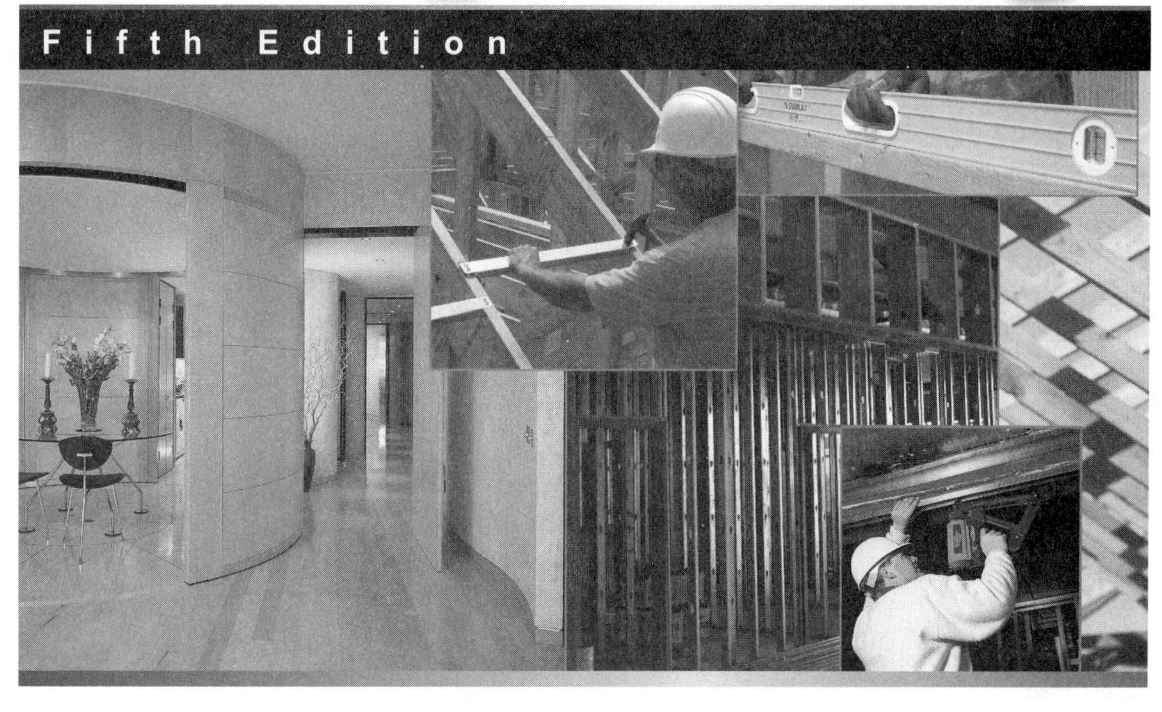

AMERICAN TECHNICAL PUBLISHERS, INC.
HOMEWOOD, ILLINOIS 60430-4600

Andris Balodis

5 6 7 8 9 – 09 – 9 8 7 6 5 4 3 2

Printed in the United States of America

ISBN 978-0-8269-0807-0

Acknowledgments

The author and publisher are grateful to the following companies and organizations for providing information, content, illustrations, and technical assistance:

BILD RenoMark™
Canada Plan Service
Canadian Centre for Occupational Health and Safety (CCOHS)
Canadian Hardwood Plywood and Veneer Association (ACCPBD/CHPVA)
Canadian Home Builders' Association (CHBA)
Canadian Lumber Standards Accreditation Board (CLSAB)
Canadian Precast/Prestressed Concrete Institute (CPCI)
Canadian Sheet Steel Building Institute (CSSBI)
Canadian Standards Association
Canadian Wood Council
Canadian Wood Truss Association (CWTA)
CANPLY, the plywood sector of CertiWood
Council of Forest Industries (COFI)
Health Canada—WHMIS
National Lumber Grades Authority (NLGA)
National Research Council Canada (NRC)
Natural Resources Canada

Special assistance provided by:
Carpenters' District Council of Ontario
Mississauga, Ontario

Table of Contents

Introduction

In the past, many textbooks and teaching resources available to Canadian learners and instructors have been predominantly (and in many cases exclusively) tailored to the practices and standards of the United States. The *Canadian Resource Supplement* provides information and resources specific to the Canadian building industry, including the following:

- Explanation of the Red Seal trade certification system
- Important differences in legislative approach
- References to Canadian building codes
- Commentary on key differences in trade practice
- Information on Canadian trade associations
- References to Canadian Internet resources
- References to Canadian material and product standards
- Metric equivalents to information given in imperial units
- Imperial to metric conversions

Resources such as Internet links and conversion tables are also found under Canadian Resources on the Interactive CD-ROM packaged with the textbook. The interactive CD-ROM included with the book features the following:

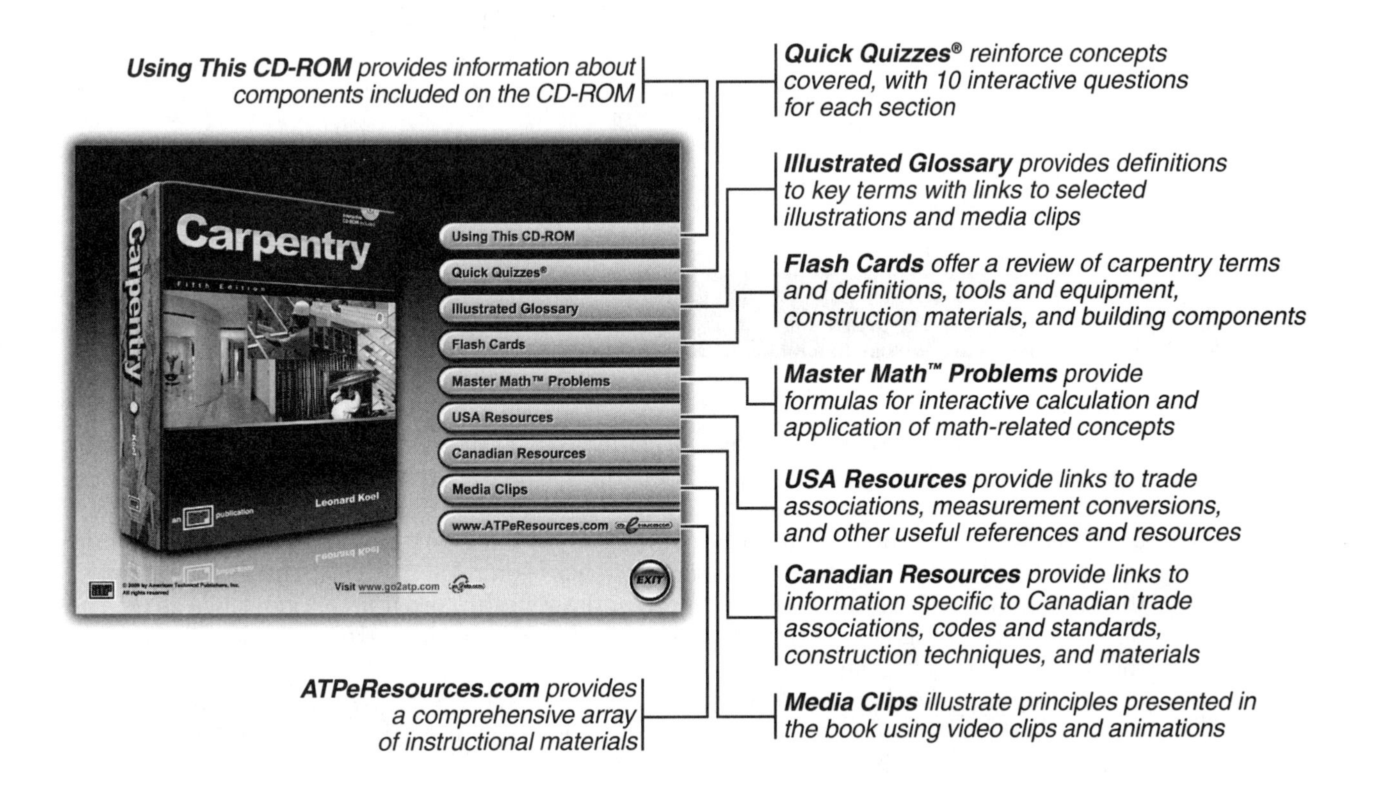

How to Use the *Supplement*

In *Carpentry*, the maple leaf icon (🍁) is located next to selected headings, figures, and appendix tables. This icon indicates that there is corresponding reference material in the *Canadian Resource Supplement*. Reference material in the *Supplement* is organized by page number in the order in which the corresponding icons appear in *Carpentry*.

In some cases, the *Supplement* contains information for Canadian users that replaces the information in the text, such as a building code reference. In other cases, the *Supplement* provides an explanation, such as a Canadian trade practice that may differ from its United States equivalent. Many of the references provide additional information, such as a reference to a web site for a Canadian-based organization not included in *Carpentry*.

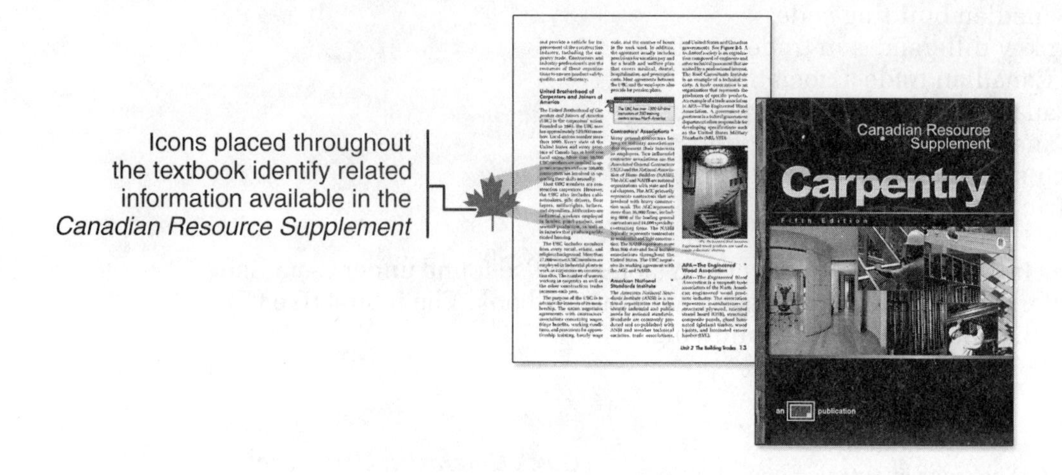

About the Author

Andris Balodis is a Red Seal certified carpenter, and a full-time professor at Conestoga College in Waterloo, Ontario, where he also is the Carpentry Apprenticeship Coordinator. In addtion to more than 10 years teaching experience, Andris has more than 20 years experience as a carpenter and design-build contractor. He has been extensively involved in curriculum development and program design. He has been responsible for the development of the Women in Skilled Trades carpentry pre-apprenticeship program at Conestoga as well as other carpentry programs and courses at the college, including extensive contributions to the Renovation Technician diploma program.

Outside of the college, Andris has volunteered with Skills Ontario, Habitat for Humanity, and various other nonprofit organizations as an organizer, carpenter, editor, and technical writer. He has also served on a number of committees devoted to carpentry curriculum development.

The Construction Industry and the Carpentry Trade in Canada

The construction industry and the carpentry trade in Canada share common practices, materials, and tools with their counterparts in the United States. Nevertheless, there are important differences in the implementation of some key aspects of building practice, and a great many more differences in the details. Legislative differences exist, and the support network provided by trade associations varies. Even the path a carpenter follows from apprentice to journeyperson status varies slightly between Canada and the United States, even though the same goal is reached.

Apprenticeship and Red Seal Trades in Canada

There are a wide variety of recognized trades in the Canadian construction industry. Some of these trades require that workers be either registered apprentices or certified journeypeople in order to work in the trade. In other trades such as carpentry, certification is optional. The advantages of becoming certified include higher wages, access to employers who require certification, and respect from those both inside and outside the trade.

Some certified trades are officially recognized only at a provincial level, and thus may be of little value to a worker who wishes to relocate to another province or territory in search of work. The nationally recognized trades in Canada, such as General Carpenter, are known as Red Seal trades, and the holders of a Red Seal are eligible to work anywhere in Canada. For some trades, the Red Seal is acquired by passing an Interprovincial Standards Examination after first qualifying at the provincial or territorial level. Other trades, such as General Carpenter, have only a single examination that is standardized across the country and automatically confers the Red Seal status.

A Red Seal trade is based on the requirements identified by the National Occupational Analysis (NOA) for the trade. This includes a list of essential skills identified by the Canadian government through consultation with industry professionals across the country. The skills that are common to all provinces and territories are selected, and the percentage of time devoted to learning each skill is determined. From this, each province develops a training standard. Training standards and the total amount of time an apprentice spends in training may vary from one region to another, but skills determined by the NOA must be included. Training can be delivered by colleges, trade unions, and private institutions recognized by the province or territory.

After successful completion of in-school training, training institutions award apprentices with a certificate of completion. A year usually passes between the completion of in-school instruction and the receipt of final certification in a trade, during which time an apprentice gains practical experience on the job. A majority of apprenticeship training is understood to take place on the job site, and requires the apprentice to participate in all aspects of the trade as defined by the NOA. Upon completion of the required on-the-job and in-school training, an apprentice may take the certification exam. The Certificate of Qualification (commonly known as the "licence") and Red Seal Status is obtained when an apprentice successfully passes the interprovincial certification exam.

Canadian Legislation and the Building Trades

In the United States, the federal government has legislative control over many areas that directly affect the construction industry, such as health and safety legislation. In Canada, the federal government often acts only as a resource and advisory body to provinces and territories, which are responsible for health and safety legislation and other legal aspects that affect construction workers. An exception to this is the Workplace Hazardous Materials Information System (WHMIS) legislation, which is national in scope. Another exception is legislation related to human rights issues.

The Canadian government also retains control over federally funded programs in the construction sector. This includes buildings constructed with Canada Mortgage and Housing Corporation (CMHC) support, the Office of Energy Efficiency R-2000 program of Natural Resources Canada, and various training incentive programs for employers and apprentices. Many of these programs also have counterparts at the provincial and territorial levels. Where indicated in the text, the *Supplement* refers to equivalent Canadian legislation.

Building Codes

The structure of building codes in Canada is similar to that in the United States. The *National Building Code of Canada (NBC)*, developed by the National Research Council (NRC), is the model code that sets the minimum standard for construction across the country. Each province or territory may issue a provincial or territorial code that contains the provisions of the *NBC*, as well as further requirements specific to the province or territory. For example, codes in Atlantic Canada may contain stricter requirements for roof coverings in recognition of wind and weather extremes encountered in those areas compared to other parts of Canada. In addition, local codes may apply in municipalities where special requirements are necessary for things such as local termite infestations or greater than average frost penetration. In some cases, code requirements are waived for special reasons. For example, in northern Canada, the normal requirements for attic venting are not applied due to the risk of airborne ice particles entering through attic vents.

If the population base of a territory or province is too small to justify issuing a specific building code, the *NBC* is followed. A similar treatment applies to fire codes. The International Codes frequently quoted in *Carpentry* are not applied in Canada.

Carpentry contains numerous specific references to building code requirements and standards. Where indicated in the text, the *Supplement* refers to equivalent Canadian building code information. In all cases, the appropriate section of the *NBC* or provincial building code should be consulted for the correct Canadian information. This is particularly true for "Unit 64—Stairway Construction." The entire unit has been reproduced in the *Supplement* to address Canadian code requirements.

Differences in Trade Practice

Due to their large geographical areas, climatic conditions of Canada and the United States can range from desert to temperate rainforest, and from humid subtropical conditions to dry tundra. Across North America, buildings may experience little wind loading or may have to resist frequent hurricane conditions. They may be resting on bedrock at or near the surface or on sand or clay with low bearing capacity. They may be located on a fault zone or on the stable Canadian Shield. Specific local building code requirements, derived from engineering calculations and experience, normally account for these factors.

Geographic conditions in the northern United States are similar to conditions in southern Canada, and building practices are likewise similar in these areas. However, there are major differences in building conditions, availability of materials, and traditional methods employed by tradespeople of northern Canada compared to those employed by tradespeople in the southern United States. Differences have evolved to suit the needs of both areas, such as the extensive use of wood siding in the north versus stucco in the south. The necessary set of materials are readily available in both respective areas, are associated with local building style, and perform well under local conditions, but neither would substitute well for the other. Builders should understand these factors and avoid introducing methods and materials that do not perform well, such as the use of cement stucco in areas where materials are subject to freeze-thaw cycle degradation.

Some differences are neutral in terms of impact. For example, Canadian homes are commonly built with full basements, while crawlspace construction is very common in the United States. In the United States, homes may use smaller dimension floor joists supported on closely spaced shallow beams resting on a number of posts. This design works well for a crawlspace used only for services. Canadian crawlspaces, by contrast, are typically constructed in the same manner as full basements, using a single large beam resting on one or two columns and supporting larger joists. Where indicated in the text, the *Supplement* refers to applicable differences in trade practice.

Trade Associations in Canada

As in the United States, the Canadian construction industry has a number of associations composed of corporations, individuals, and in some cases government representatives. These associations perform a number of functions that commonly include self-regulation, product and service promotion, information and resource sharing, and representation of its members in a united voiced. Most American trade associations have a Canadian equivalent, though there are some trade associations in both countries that have no equal counterpart. In some cases, a single association represents interests in both countries.

Trade associations usually can provide useful information to builders, such as lists of member suppliers, specific product information, and installation or usage guides. Many, such as the Canadian Home Builders Association (CHBA), produce valuable publications. While some of this information is only available to members or must be paid for, much of it is free and readily available either on the Internet or as a printed resource. Where indicated in the text, the *Supplement* refers to equivalent Canadian trade associations.

Internet Resources

The Internet is a global resource, so to some extent it is misleading to refer to "Canadian" Internet resources. However, Canadian carpenters should learn to recognize resources that are based in Canada and contain standards or legal references that are in effect in Canada. An example of this is workplace safety legislation.

When searching for information, it is often useful to restrict the search engine to "Canada only" by clicking on the appropriate button. An Internet user can also try substituting .ca for .com in the URL.

The *Supplement* contains references to Internet resources and web sites useful to Canadian carpenters. There are many more useful Canadian resources and related Internet links on ATPeResources.com and on the interactive CD-ROM included with *Carpentry*.

Material and Product Standards

Many American standards have been adopted globally, such as standards set by ASTM International. These and other standards are widely used in Canada and often referenced in building codes and specifications. Other standards and grading systems have been developed by trade associations, and products that are regularly sold in both countries may bear markings from both Canadian and American associations. An example of this is plywood, which may bear markings from APA—The Engineered Wood Association, the Council of Forest Industries (COFI), or both. Some American organizations, such as Underwriters Laboratories (UL), have a Canadian division. In addition, Canada has its own agencies that set standards and monitor quality, such as the Canadian Standards Association (CSA), which sets Canadian standards, and CSA International, which tests and certifies that products meet those standards. Where indicated in the text, the *Supplement* refers to equivalent Canadian standards.

CSA Certification Marks

FOR CANADA

FOR CANADA AND THE U.S.

MEETS CANADIAN STANDARDS FOR APPLIANCES USING GAS OR PETROLEUM

CERTIFIED

FOR CANADIAN GAS PRODUCTS

Metric Measurements

In many cases, *Carpentry* provides specific measurements or quantities in imperial units. The construction industry in Canada, in large part, uses the imperial system. The exception to this is work funded by the federal government, which is dimensioned in metric units. For this reason, much of the content provided in imperial units is equally applicable to Canadian construction. Where a reference to the *Supplement* has been included, it is typically to offer an explanation, such as the sizing system for steel beams, or because the measurements provided are not consistent with Canadian codes.

Most building materials in Canada are the same size as their counterparts in the United States, but the sizes have been "soft converted," meaning that measurements are given in metric units in the *NBC* and in government documents. Thus, a 2 × 4 is the same size on both sides of the border, but may be referred to in Canada as a 38 × 89. For lumber available in stock, length is typically given in imperial units. If metric length is desired, it may be ordered to size.

Panel products normally measure 4′ × 8′, or other standard imperial sizes, and are usually referred to as such, but thicknesses have been standardized to metric units because the difference is small and has little practical effect on construction. Panels are available in true metric width and length if required.

One notable exception to this is asphalt roof shingles, which are usually 3′ × 1′ in the United States, but are made and sold in Canada sized 1 metre × ⅓ metre.

Where indicated in the text, the *Supplement* refers to any instance where the difference in measurement systems has a significant impact. There are also conversion tables provided in the *Supplement*.

Page 13
Contractors' Associations

The Canadian Home Builders' Association (CHBA) provides resources and support for builders and renovators, as well as information and advice for home owners. In order to maintain high industry standards, the CHBA has also developed a code of ethics for its member builders and renovators. More information is available at **www.chba.ca** and **www.renomark.ca.**

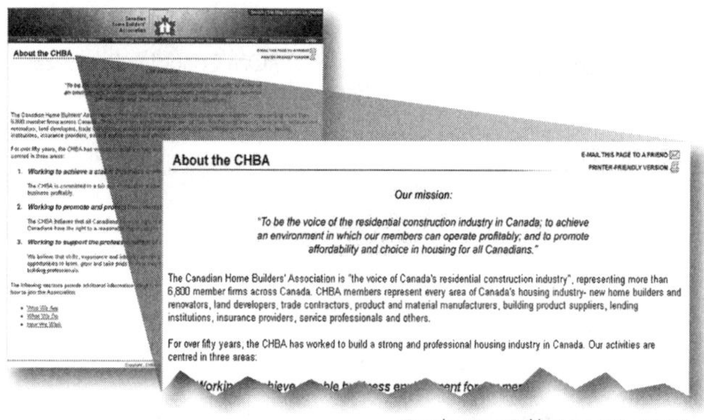

Canadian Home Builders' Association (CHBA)

See Carpentry CD-ROM Canadian Resources–Canadian Home Builders' Association (CHBA)

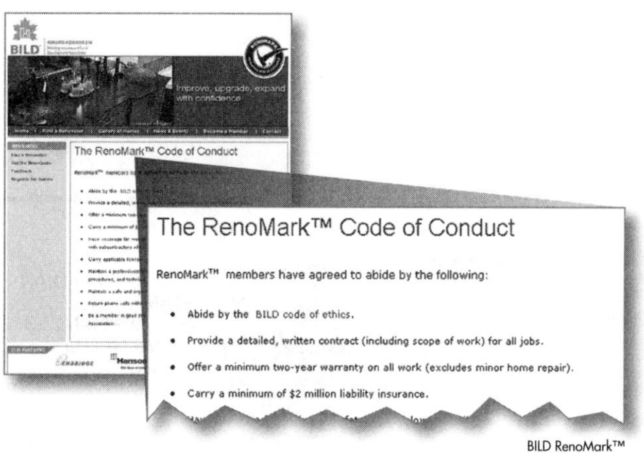

BILD RenoMark™

See Carpentry CD-ROM Canadian Resources–RenoMark™

Page 13
National Research Council Canada (NRC) and the Canadian Standards Association (CSA)

The National Research Council Canada (NRC) is a federal government organization for research and development. More information is available at **www.nrc-cnrc.gc.ca.** The CSA International tests products for compliance with national and international standards and issues product certifications. More information is available at **www.csa-international.org.**

See *Carpentry* CD-ROM Canadian Resources–National Research Council Canada (NRC)

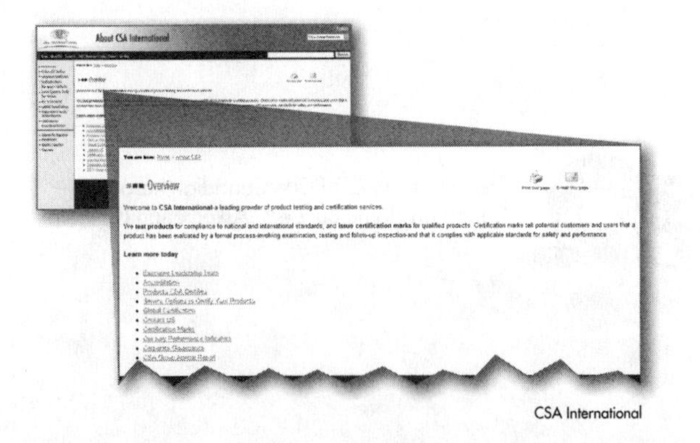

CSA International

See *Carpentry* CD-ROM Canadian Resources–CSA International

CSA Certification Marks

FOR CANADA

FOR CANADA AND THE U.S.

FOR CANADIAN GAS PRODUCTS

Page 13
Council of Forest Industries (COFI)

The Council of Forest Industries (COFI) represents forest product producers in British Columbia and is used across Canada as a resource similar to APA—The Engineered Wood Association. More information is available at **www.cofi.org.**

APA—The Engineered Wood Association

APA—The Engineered Wood Association is a nonprofit trade association of the North American engineered wood products industry. The association

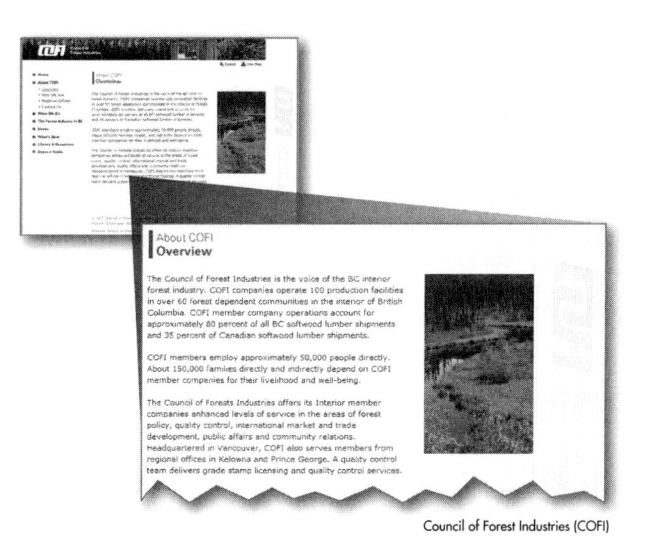

Council of Forest Industries (COFI)

See *Carpentry* CD-ROM Canadian Resources–Council of Forest Industries (COFI)

Page 15
Canadian Wood Council (CWC)

The Canadian Wood Council (CWC) is a national association representing manufacturers of wood products. They promote wood products through education and product information services. More information is available at **www.cwc.ca.**

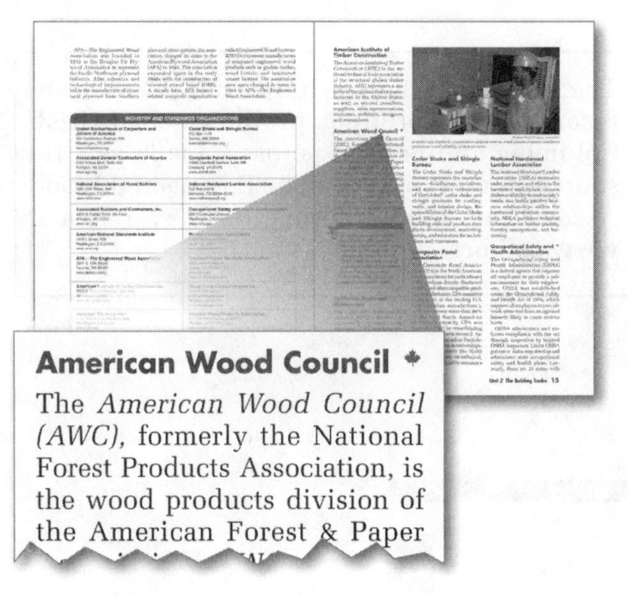
American Wood Council

The *American Wood Council (AWC)*, formerly the National Forest Products Association, is the wood products division of the American Forest & Paper

CWC Home › About CWC

Who we are?

The Canadian Wood Council CWC is the national association representing manufacturers of Canadian wood products used in construction. Through its member Associations, the Council represents those manufacturers.

How we add value

The Canadian Wood Council enables the selling of Canadian wood products through programs and services focused on creating market access and demand. CWC increases the use of Canadian wood by assuring their broad regulatory acceptance. The Council ensures that North American construction can make maximum use of Canadian wood products with guidelines, tools and education. With wood being increasingly sought after for its high performance and green building attributes, the Council's mandate is more than ever critical as it works with regulators, design and building professionals to specify wood in today's modern building systems. Not only does the Council contribute to the financial well being of Canada' wood industry, it also ensures the well being of communities where the industry operates. The Canadian wood industry is vital to Canada's economy which includes generating a significant number of direct and indirect jobs from coast to coast!

Mission and Vision

See *Carpentry* CD-ROM Canadian Resources–Canadian Wood Council (CWC)

Canadian Wood Council (CWC)

Page 15
Canadian Centre for Occupational Health and Safety (CCOHS)

The Canadian national organization for workplace health and safety is the Canadian Centre for Occupational Health and Safety (CCOHS). Most workplace safety legislation in Canada is enacted and enforced by each province or territory. More information is available at **www.ccohs.ca.**

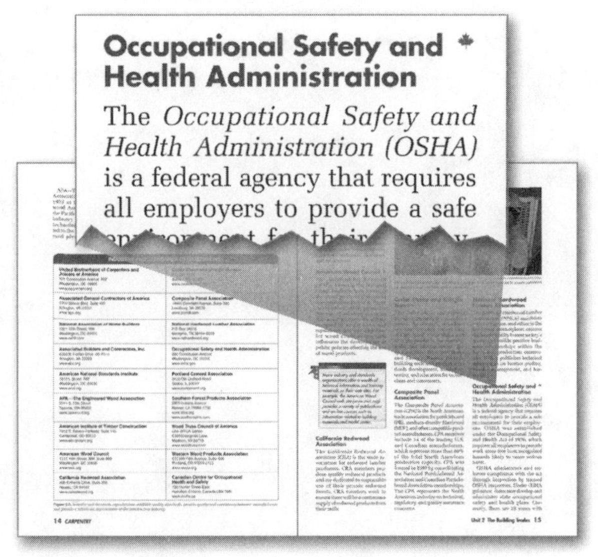
Occupational Safety and Health Administration

The *Occupational Safety and Health Administration (OSHA)* is a federal agency that requires all employers to provide a safe

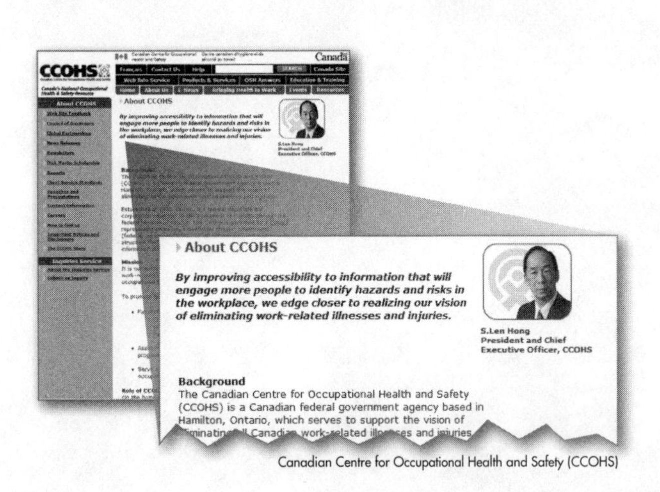
About CCOHS

By improving accessibility to information that will engage more people to identify hazards and risks in the workplace, we edge closer to realizing our vision of eliminating work-related illnesses and injuries.

S.Len Hong
President and Chief Executive Officer, CCOHS

Background

The Canadian Centre for Occupational Health and Safety (CCOHS) is a Canadian federal government agency based in Hamilton, Ontario, which serves to support the vision of

See *Carpentry* CD-ROM Canadian Resources–Canadian Centre for Occupational Health and Safety

Canadian Centre for Occupational Health and Safety (CCOHS)

Page 16
Apprenticeship Programs

The nationally recognized trades in Canada, such as General Carpenter, are known as Red Seal trades, and the holders of a Red Seal are eligible to work anywhere in Canada. A Red Seal trade is based on the requirements identified by the National Occupational Analysis (NOA) for the trade. More information is available at **www.red-seal.ca**

See Carpentry CD-ROM Canadian Resources–Interprovincial Standards Red Seal Program

Page 33
Softwood Grading Systems

The National Lumber Grades Authority (NLGA) is responsible for writing, interpreting, and maintaining Canadian lumber grading rules and standards. More information is available at **www.nlga.org.** The Canadian Lumber Standards Accreditation Board (CLSAB) is an accreditation body for lumber grading agencies. More information is available at **www.clsab.ca.**

The Canadian Wood Council publishes information on Canadian lumber grade stamps, grades and uses of Canadian dimension lumber, strip flooring grades, and softwood grades for interior finish work in *Wood Reference Handbook*.

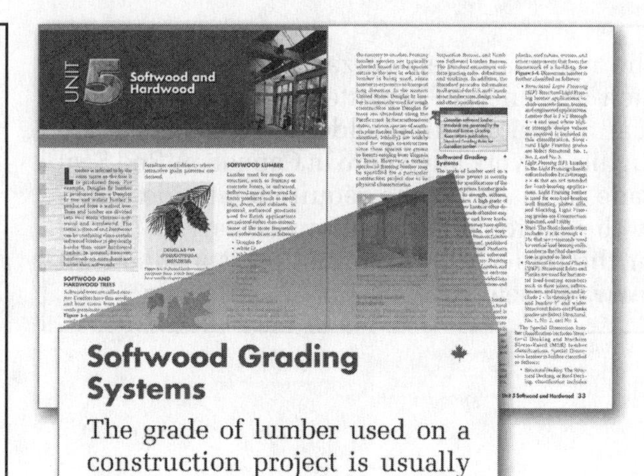

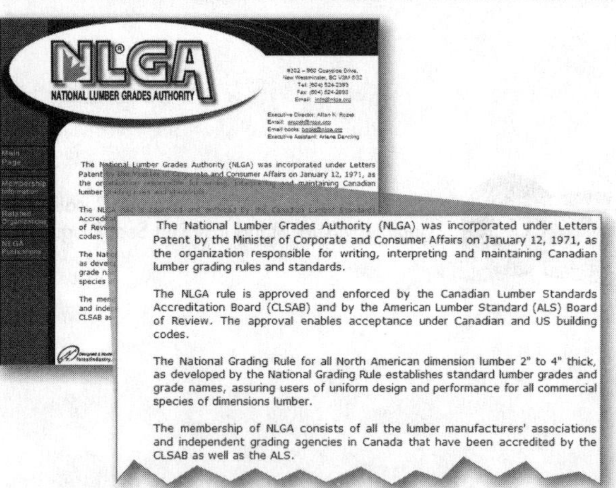

National Lumber Grades Authority (NLGA)

See Carpentry CD-ROM Canadian Resources–National Lumber Grades Authority (NLGA)

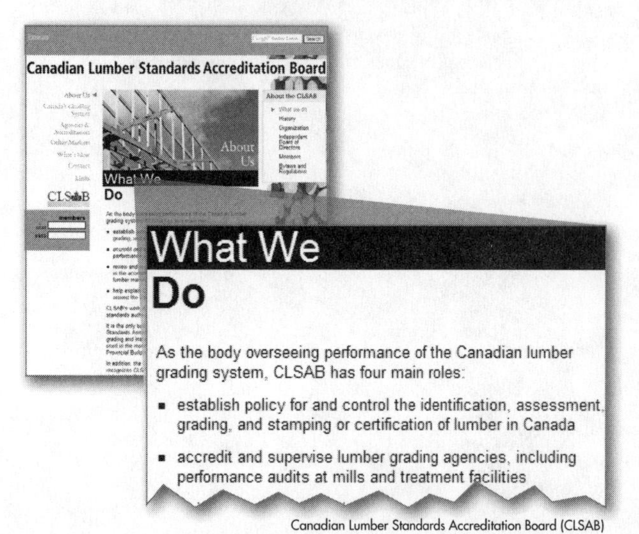

Canadian Lumber Standards Accreditation Board (CLSAB)

See *Carpentry* CD-ROM Canadian Resources–Canadian Lumber Standards Accreditation Board (CLSAB)

Canadian Lumber Grade Stamps

A.F.P.A.® 00 — Mill designation
(Grading agency)
S–P–F — Species group
S-DRY — Moisture content
No. 1 — Assigned Grade

Alberta Forest Products Association
11738 Kingsway Avenue, Suite 200
Edmonton, AB T5G 0X5
780-452-2841
www.abforestprod.org
(Approved to supervise fingerjoining and machine stress-rated lumber)

A.F.P.A.® 00
S–P–F
S-DRY
No. 1

Association des manufacturiers de bois de sciage du Québec (Québec Lumber Manufacturers' Association)
5055 boul. Hamel ouest, bureau 200
Québec, QC G2E 2G6
418-872-5610
www.sciage-lumber.qc.ca
(Approved to supervise fingerjoining and machine stress-rated lumber)

Canadian Lumbermen's Association
27 Goulburn Avenue
Ottawa, ON K1N 8C7
613-233-6205
www.cla-ca.ca
(Approved to supervise fingerjoining and machine stress-rated lumber)

C L® A
S-P-F
100
No. 1
S-GRN.

Cariboo Lumber Manufacturers' Association
197 Second Avenue North, Suite 205
Williams Lake, BC V2G 1Z5
250-392-7778
www.clma.com
(Approved to supervise fingerjoining and machine stress-rated lumber)

CLMA 1 1 S-GRN 1
D FIR (N)

Excerpt from *Wood Reference Handbook,* ©1991 by Canadian Wood Council

Canadian Dimension Lumber—Grades and Uses

Grade Category	Grades	Common Grade Mix	Principal Uses
Structural Light-Framing 38 to 89mm (2" to 4" nom.) thick 38 to 89mm (2" to 4" nom.) wide	Select Structural No.1 No.2 No.3	No.2 and Better	Used for engineering applications such as for trusses, lintels, rafters and joists in the smaller dimensions.
Structural Joists and Planks 38 to 89mm (2" to 4" nom.) thick 114mm (5" nom.) or wider	Select Structural No.1 No.2 No.3	No.2 and Better	Used for engineering applications such as for trusses, lintels, rafters, and joists in the dimensions greater than 114mm (5" nom.).
Light Framing 38 to 89mm (2" to 4" nom.) thick 38 to 89mm (2" to 4" nom.) wide	Construction Standard Utility	Standard and Better (Std. & Btr.)	Used for general framing where high strength values are not required such as for plates, sills, and blocking.
Stud 38 to 89mm (2" to 4" nom.) thick 38mm (2" nom.) or wider 3m (10') or less in length	Stud Economy Stud		Made principally for use in walls. Stud grade is suitable for bearing wall applications. Economy grade is suitable for temporary applications.

Notes:
1. Grades may be bundled individually or they may be individually stamped but they must be grouped together with the engineering properties dictated by the lowest strength grade in the bundle.
2. The common grade mix shown is the most economical blending of strength for most applications where appearance is not a factor and average strength is acceptable.
3. Except for economy grade, all grades are stress graded which means specified strengths have been assigned and span tables calculated. Economy and utility grades are suited for temporary construction or for applications where strength and appearance are not important.
4. Construction, Standard, Stud, and No.3 grades should be used in designs that are composed of 3 or more essentially parallel members (load sharing) spaced at 610mm (2') centres or less.
5. Strength properties and appearance are best in the premium grades such as Select Structural.

Canadian Strip Flooring Grades

Species	Grade	Description of Grade
Birch and Maple	Clear Grade Birch and First Grade Maple	Face free of defects, mild discolourations and pin knots less than 3mm (1/8") diameter admitted. Natural colour variation admitted. 600 to 900mm (2' to 3') bundles not to exceed 30 percent of shipment. Minimum length bundle 600mm (2').
	Select Grade Birch and Second Grade Maple	Tight sound knots admitted. Slight dressing imperfections admitted. 450 to 900mm (1-1/2' to 3') bundles not to exceed 45 percent of shipment. Minimum length bundle 450mm (1-1/2').
	Third Grade Birch and Factory Grade Maple	Sound defects admitted. Some cutting of pieces admitted. 300 to 900mm (1' to 3') bundles not to exceed 60 percent of shipment. Minimum length bundle 300mm (1').
Plain Sawn Oak	Clear Grade	Face free of defects. 9.5mm (3/8") sapwood admitted. Natural colour variation admitted. Bundles shorter than 1200mm (4') not to exceed 45 percent of shipment. Minimum length bundle 600mm (2').
	Select Grade	Pinworm holes or small knots admitted. Slight dressing imperfections admitted. 450 to 900mm (1-1/2' to 3') bundles not to exceed 50 percent of shipment. Minimum length bundle 300mm (1').
	No.1 Common	Sound defects admitted. Cutting of pieces not admitted. 300 to 900mm (1' to 3') bundles not to exceed 50 percent of shipment. Minimum length bundle 300mm (1').
	No.2 Common	Admits all defects. A mill run grade suitable for construction where appearance is not the requisite.
Quarter-Sawn Oak	First Grade	Face free of defects. 9.5mm (3/8") sapwood admitted. Natural colour variation admitted. Bundles shorter than 1050mm (3-1/2') not to exceed 25 percent of shipment. Minimum length bundle 600mm (2').
	Sap Clear First Grade	Sapwood, pinworm holes or small knots admitted. Slight dressing imperfections admitted. Bundles shorter than 1050mm (3-1/2') not to exceed 25 percent of shipment. Minimum length bundle 600mm (2').
	Second	Sapwood, pinworm holes, slight dressing imperfections and small knots admitted. 450 to 900mm (1-1/2' to 3') bundles not to exceed 40 percent of shipment. Minimum length bundle 450mm (1-1/2').

Note:
Combination grades, which contain a mixture of grades of a given species, extend the range of sizes and prices available. Specifiers should ask a manufacturer for details or inquire at the Canadian Lumbermen's Association (see Information Sources).

Appearance Characteristics of Softwoods used for Interior Finish

Species	Characteristics
Western red cedar	Reddish brown, close even grain, annual layers easily distinguishable, relatively soft.
Douglas fir	Heartwood bright red to yellow, sapwood nearly white, relatively free of resins, pronounced variable rings, hard surface.
Pacific coast hemlock	Heartwood reddish brown, sapwood nearly white, close straight grained, relatively free of resins, hard surface.
Lodgepole pine	Heartwood light brown to red, sapwood nearly white, close straight grained, relatively soft.
Sitka spruce	Heartwood light reddish brown, sapwood nearly white, medium-coarse grain, fairly durable.
Eastern white pine	Heartwood cream white, sapwood mainly white, ring width variable close straight grain, fairly resinous, relatively soft.
Jack pine	Heartwood bright red, sapwood yellow to white ring width variable pronounced saps, fairly resinous, fairly durable.
Red pine	Heartwood bright red, sapwood yellow to white ring width variable pronounced saps, fairly resinous, fairly durable.
Eastern white cedar	Heartwood light reddish brown, sapwood nearly white, close straight grained, relatively soft.

Suggested Softwood Grades for Interior Finish Work

Species	Grade Category	Grades	Characteristics
All species except Western red cedar	K.D. (kiln dry) Finish	C and Better	Small imperfections, suitable for clear finishing.
		D	Larger imperfections, suitable for painted surfaces.
	K.D. Ceiling & Siding	C and Better D	Same as for K.D. Finish. Same as for K.D. Finish.
	K.D. Casing & Base	C and Better D	Same as for K.D. Finish. Same as for K.D. Finish.
	Industrial Clears	B and Better	Supreme grade, clear on 1 face.
		C	Small imperfections.
		D	Utility grade for surfaces to be painted.
Eastern white pine and red pine	Selects	B and Better	Many pieces are completely clear.
		C	Small imperfections, suitable for clear finishing.
		D	More imperfections, suited for less exacting applications.
Western red cedar	Finish, Panelling, and Drop Siding K.D.	Clear Heart	Many pieces are completely clear.
		A	Small imperfections, suited for clear finishes.
		B	More imperfections, short lengths defect free.
	Industrial Clears	B and Better	Supreme grade, clear on 1 face.
		C	Small imperfections.
		D	Utility grade for surfaces to be painted .
	Tight Knotted Panelling and Siding	Select	Best quality where knotty appearances desired.
		Utility	Some loose knots permitted.

Engineered Wood Panels

Page 42
Engineered Wood Panels

Canadian builders use guidelines developed by APA—The Engineered Wood Association as well as by COFI and Canadian Plywood (CANPLY).

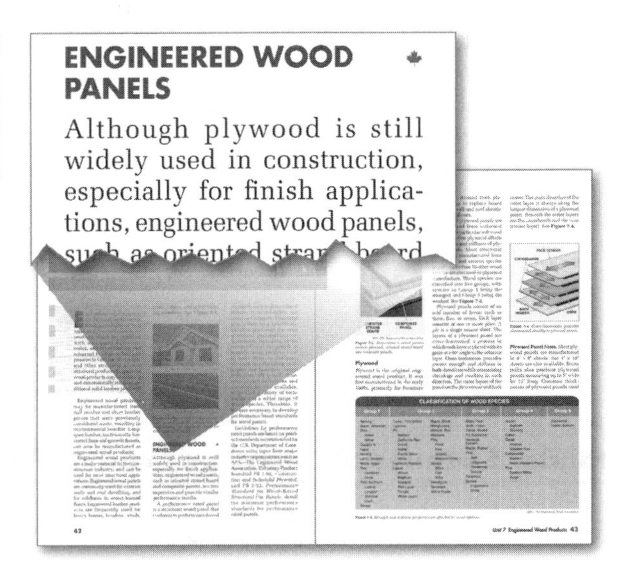

Page 44
Overlaid Plywood

Canadian Plywood (CANPLY) provides information on plywood for concrete forms. More information is available at **www.canply.org/english/products/cofiform.htm.**

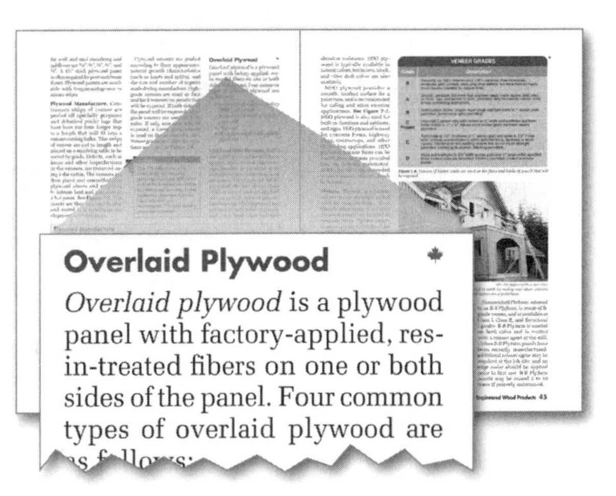

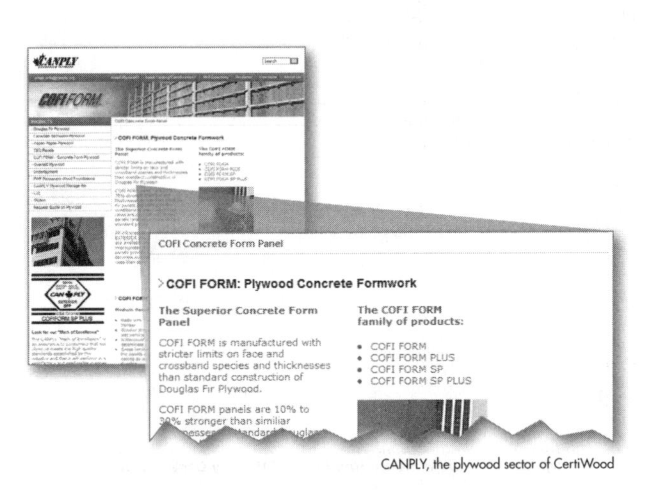

CANPLY, the plywood sector of CertiWood

See *Carpentry* CD-ROM Canadian Resources—CANPLY, the plywood sector of CertiWood

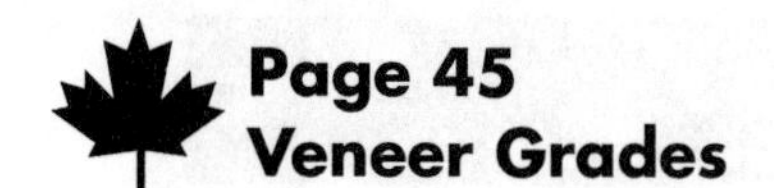

Page 45
Veneer Grades

Canadian Plywood (CANPLY) provides information on Canadian sheathing plywood. More information is available at **www.canply.org/english/products/csp/csp_unsanded_shg.htm.** The Canadian Wood Council publishes information on veneer grades for exterior plywood in *Wood Reference Handbook.*

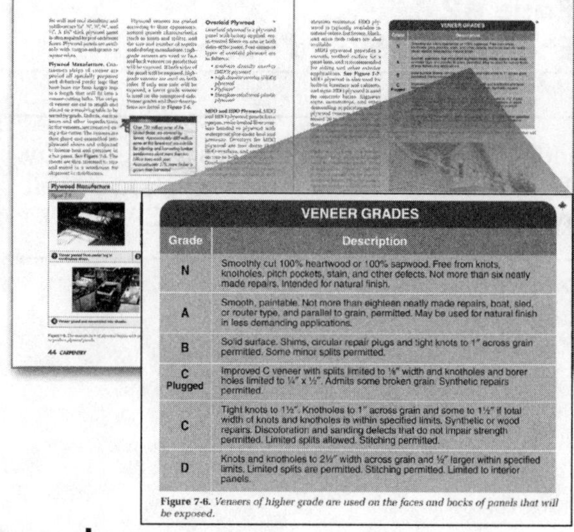

VENEER GRADES

Grade	Description
N	Smoothly cut 100% heartwood or 100% sapwood. Free from knots, knotholes, pitch pockets, stain, and other defects. Not more than six neatly made repairs. Intended for natural finish.
A	Smooth, paintable. Not more than eighteen neatly made repairs, boat, sled, or router type, and parallel to grain, permitted. May be used for natural finish in less demanding applications.
B	Solid surface. Shims, circular repair plugs and tight knots to 1" across grain permitted. Some minor splits permitted.
C Plugged	Improved C veneer with splits limited to 1/8" width and knotholes and borer holes limited to 1/4" x 1/2". Admits some broken grain. Synthetic repairs permitted.
C	Tight knots to 1 1/2". Knotholes to 1" across grain and some to 1 1/2" if total width of knots and knotholes is within specified limits. Synthetic or wood repairs. Discoloration and sanding defects that do not impair strength permitted. Limited splits allowed. Stitching permitted.
D	Knots and knotholes to 2 1/2" width across grain and 1/2" larger within specified limits. Limited splits are permitted. Stitching permitted. Limited to interior panels.

Figure 7-6. *Veneers of higher grade are used on the faces and backs of panels that will be exposed.*

Standard Grades of Canadian Exterior Plywood

Grade	Governing Canadian Standard	Individual Veneer Grades: Face	Inner Plies	Back	Characteristics	Typical Applications
Good Two Sides (G2S)	CSA O121 (DFP)	A	C	A	Sanded. Best appearance both faces. May contain neat wood patches, inlays or synthetic patching material.	Used where appearance of both sides is important. Furniture, cabinet doors, partitions, shelving, and concrete formwork.
Good One Side (G1S)	CSA O121 (DFP)	A	C	C	Sanded. Best appearance one side only. May contain neat wood patches, inlays or synthetic patching material.	Used where appearance on one side is important. Furniture, cabinet doors, partitions, shelving, and concrete formwork.
Select-Tight Face (SEL TF)	CSA O121 (DFP) or CSA O151 (CSP)	B	C	C	Unsanded. Permissible face openings filled. May be light sanded to clean and size patches.	Used where appearance on one side is important. Furniture, cabinet doors, partitions, shelving, and concrete formwork.
Select (SELECT)	CSA O121 (DFP) or CSA O151 (CSP)	B	C	C	Unsanded. Uniform surface with minor open splits. May be cleaned and sized.	Underlayment, combined subfloor and underlayment, and sheathing.
Sheathing (SHG)	CSA O121 (DFP) or CSA O151 (CSP)	C	C	C	Unsanded. Face may contain limited size knots and other defects.	Roof, wall, and floor sheathing.

Notes:
1. Permissible openings filled with wood patches or putty.
2. For information on specialty plywoods, refer to Section 7.
3. All grades are bonded with waterproof phenolic glue.
4. Veneer grades: A: highest grade; B: medium grade; and C: low grade.

Reproduced in part with the permission of COFI.

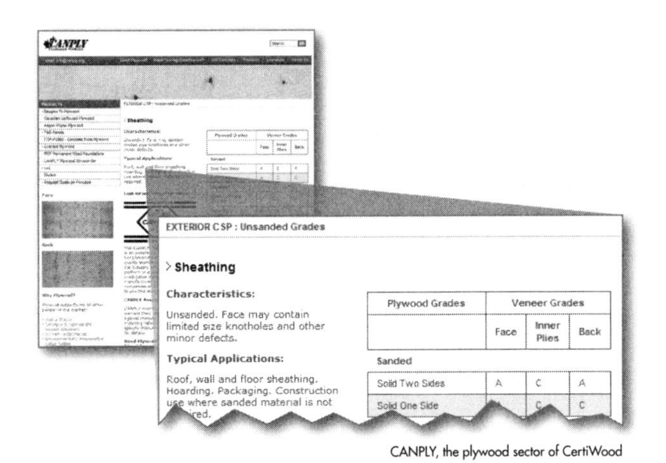

CANPLY, the plywood sector of CertiWood

See Carpentry CD-ROM Canadian Resources – CANPLY, the plywood sector of CertiWood

Page 49 Panel Trademarks

The Canadian Wood Council publishes information on panel trademarks for Canadian plywood in *Wood Reference Handbook*.

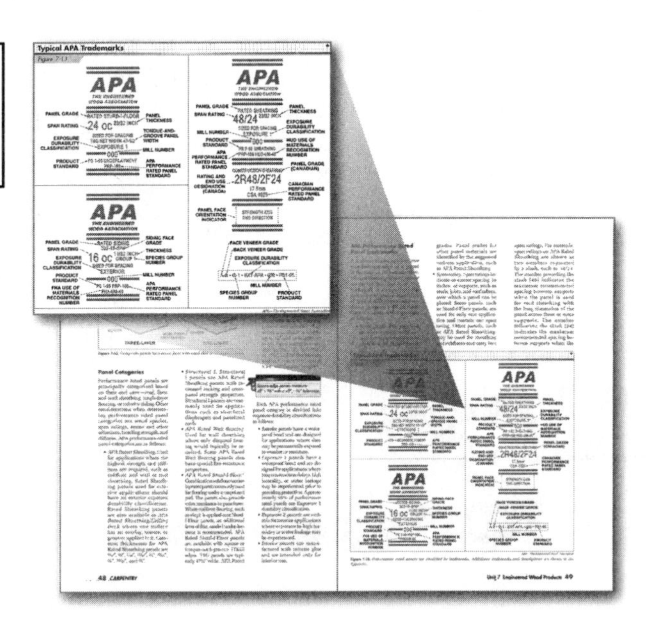

Sample Panel Marks for Canadian Plywood

Face Stamp on CANPLY EXTERIOR Plywood (Unsanded grades)

Indicates that this product is manufactured under CANPLY's Quality Certification Program.

Licensed mill number of the Canadian Plywood Association member

Indicates that the plywood has been manufactured by a member of the Canadian Plywood Association

Indicates a completely waterproof glue bond

Indicates species designation: DFP (Douglas Fir plywood), CSP (Canadian Softwood plywood) or Poplar plywood

Indicates the CSA standard governing manufacture

*CSP, DFP, or POPLAR
**CSA 0151M, CSA 0121M or CSA 0153M

Edge Stamp on CANPLY EXTERIOR Plywood (Sanded and Unsanded grades)

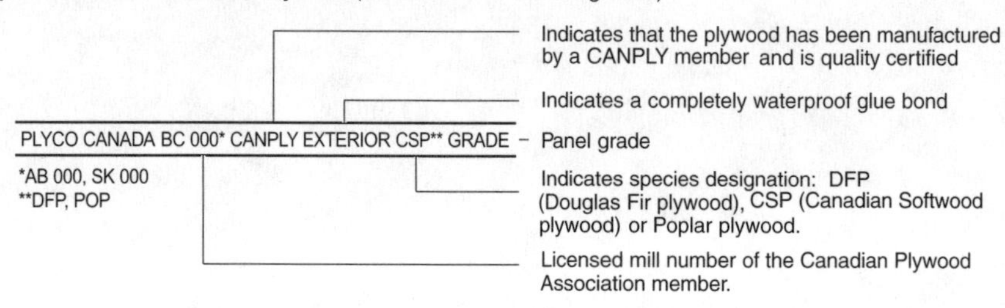

Indicates that the plywood has been manufactured by a CANPLY member and is quality certified

Indicates a completely waterproof glue bond

Panel grade

Indicates species designation: DFP (Douglas Fir plywood), CSP (Canadian Softwood plywood) or Poplar plywood.

Licensed mill number of the Canadian Plywood Association member.

*AB 000, SK 000
**DFP, POP

Excerpt from *Wood Reference Handbook*, ©1991 by Canadian Wood Council

Page 51
Hardwood Plywood Panel Grades

The Canadian Hardwood Plywood and Veneer Association (CHPVA) is a national association representing veneer and hardwood plywood manufacturers. More information is available at **www.chpva.ca**

HARDWOOD PLYWOOD PANEL GRADES

Grade	Description
A-1	Smooth face veneer running full panel length. Book-matched veneer edges if more than one veneer is used on face. May contain small burls and pin knots or small repairs. Sound veneer back, but may not be color matched.
B-2	Sound face veneer, but may have small burls, pin knots, and repairs. If face veneer consists of more than one piece, color and grain matching required. Solid back veneer with repaired voids.
C-2	Sound face veneer with more small burls, pin knots, and other growth characteristics. Sound repaired back.
C-3	Sound, repaired face. Back veneer contains more and larger repairs.
D-3	Economy-grade panel. Natural growth characteristics allowed on face and back, but no open areas allowed.

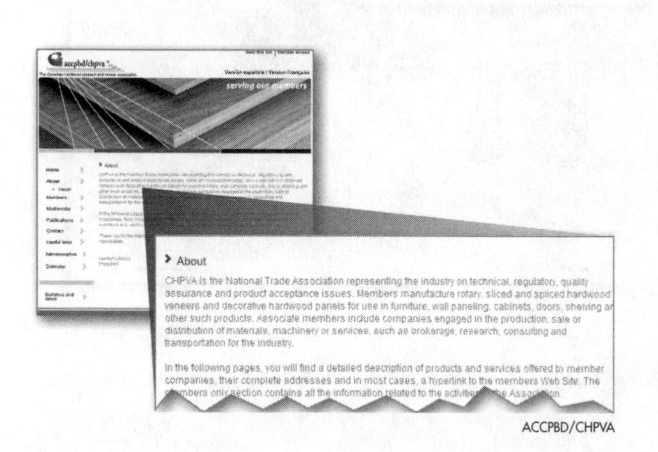

› About

CHPVA is the National Trade Association representing the industry on technical, regulatory, quality assurance and product acceptance issues. Members manufacture rotary, sliced and spliced hardwood veneers and decorative hardwood panels for use in furniture, wall paneling, cabinets, doors, shelving and other such products. Associate members include companies engaged in the production, sale or distribution of materials, machinery or services, such as brokerage, research, consulting and transportation for the industry.

In the following pages, you will find a detailed description of products and services offered by member companies, their complete addresses and in most cases, a hyperlink to the members Web Site. The members only section contains all the information related to the activities of the Association.

ACCPBD/CHPVA

See Carpentry CD-ROM Canadian Resources–Canadian Hardwood Plywood and Veneer Association (ACCPBD/CHPVA)

Grades of Canadian Hardwood Plywood

Canadian hardwood plywoods are manufactured and graded in accordance with the Canadian Hardwood Plywood Association's *Official Rules for Canadian Hardwood Plywood.*

Veneers for hardwood plywood are graded based on minimum characteristics for face and brick veneers. The *face* is the better side of a panel where the outer sides are different, and either side of a panel where the outer sides are the same. Face grades are designated AA, A, B, C, D, and E (AA is the best quality and E is the lowest).

The *back* is the side of the panel with the lower grade where the outer sides are different. Back grades are designated 1, 2, 3, and 4 (1 is the best and 4 is the lowest).

The characteristics of face grades differ slightly depending on the species and the method of veneer cutting used. In general, the better quality face grades have the following characteristics and uses:

- AA: slight colour variation, only slight imperfections permitted, highest quality, used for high quality cabinetry and architectural woodwork where a superior appearance is required.
- A: slight colour variation, sight imperfections and knots to about 4.8 mm (3/16″) permitted, used for good quality cabinetry and architectural woodwork.
- B: some colour variation, slight imperfections and knots to about 4.8 mm (3/16″) permitted, used for cabinetry and woodwork where the product is less visually exposed or where an opaque finish is to be used.

Excerpt from *Wood Reference Handbook,* pp. 419-420. ©1991 by Canadian Wood Council.

Personal Protective Equipment

Page 182
Health and Safety Legislation

Canadian health and safety legislation varies and may not require employers to purchase the required personal protective equipment.

In late 2007, OSHA announced a final rule on employer-paid personal protective equipment (PPE).* The rule provides that all PPE must be purchased by the employer when used by the worker to comply with one of

Hazardous Materials

Page 188
Hazardous Materials

Health Canada is a department of the federal government responsible for helping Canadians maintain and improve health. Health Canada provides information on the Workplace Hazardous Materials Information System (WHMIS) in Canada. Adherence to the requirements of the WHMIS is a requirement for all Canadian employers. Under WHMIS legislation, manufacturers must make key information available on their products, such as safe handling procedures, toxological properties, and first aid measures. The employer is responsible for ensuring that this information is available to workers and for providing training. The worker is required to attend such training and to follow safe workplace procedures. More information is available at **www.hc-sc.gc.ca/ewh-semt/occup-travail/whmis-simdut/index-eng.php.**

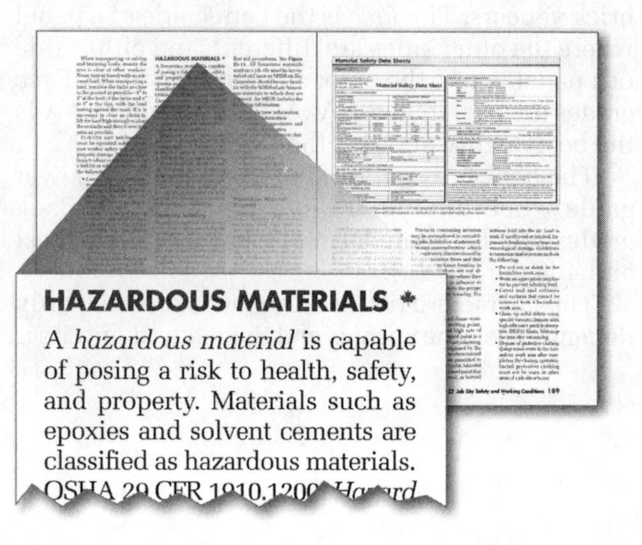

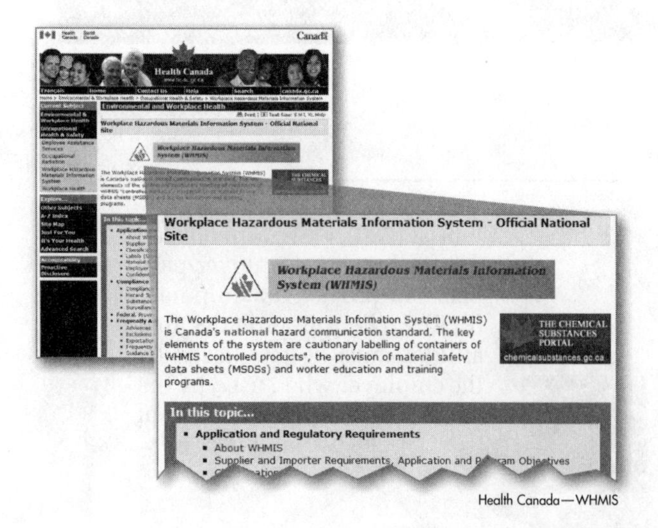

Health Canada—WHMIS

See *Carpentry* CD-ROM Canadian Resources–Health Canada—WHMIS

Excavations

Page 191
Sloping and Benching

In Canada, soil type classifications vary from province to province. As a result, requirements for sloping or benching may differ, and provincial regulations should always be consulted.

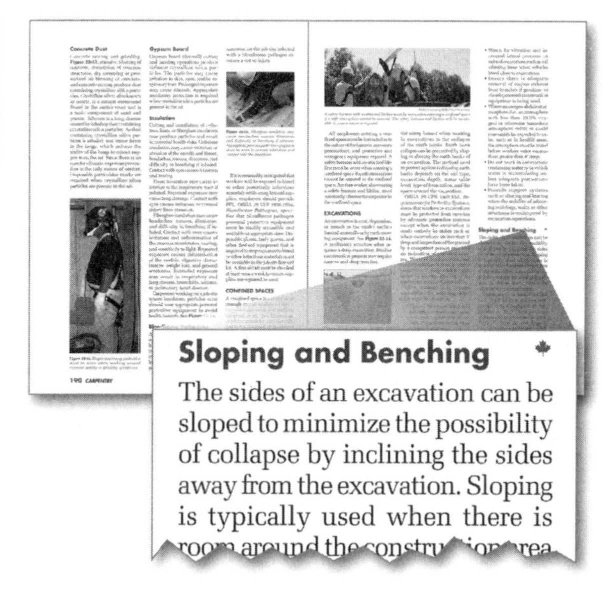

Page 192
Soil Types and Conditions

Soil types and requirements for sloping or benching may be different in Canada from those illustrated in Figure 22-15.

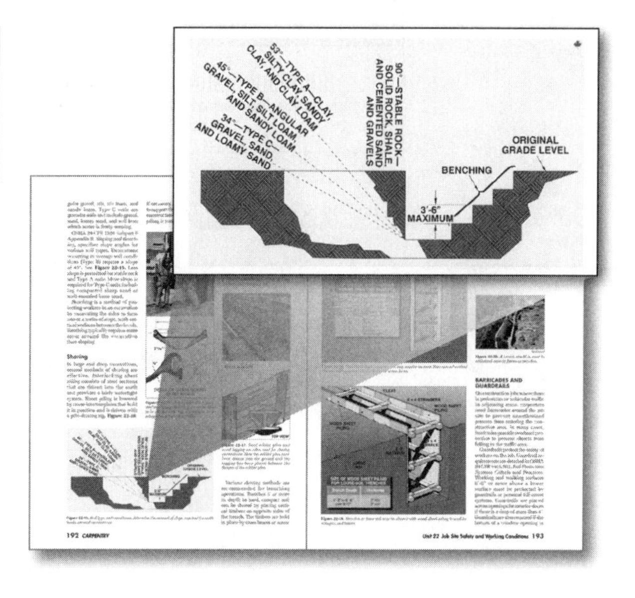

Page 207
Reading an Architect's Scale

Metric scale rules work on a simple numerical ratio. Thus, a scale identified as 1:200 has a ratio of 200 units on a blueprint to 1 unit on the rule. Unless otherwise indicated, the units used in metric prints are millimetres, so at 1:200, each scaled unit is 200 millimetres or 0.20 metre.

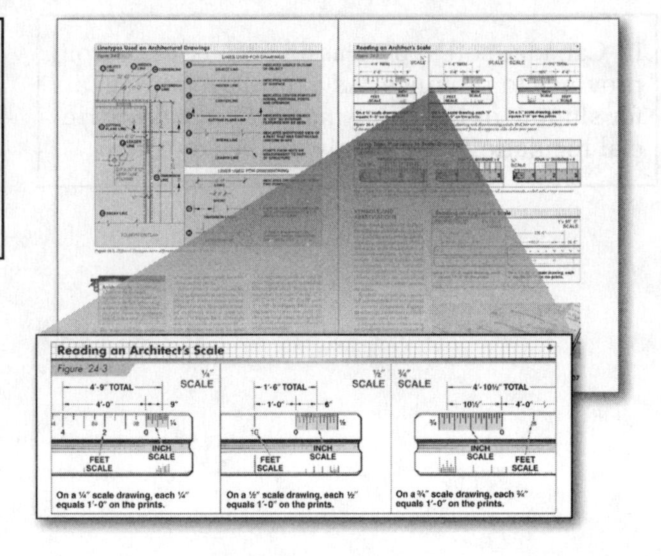

Metric Scale Rule

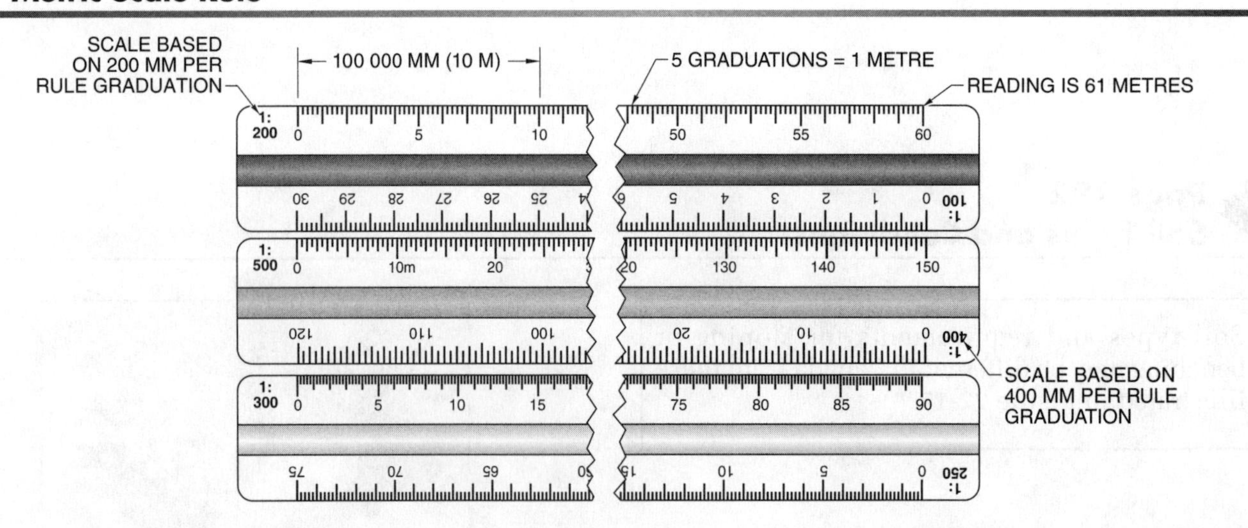

Metric scale rules use decimal system measurements.

Building Codes

Page 250
Provincial and Local Codes

The Canadian practice with respect to building codes is similar to the American practice. The *National Building Code of Canada* acts as a model code on which provincial codes are based, and municipalities may have their own additional requirements. In the absence of a provincial code, the national code is used.

Builder's Levels

Page 261
Architect's and Engineer's Rods

There are many types of levelling rods. Some rods are graduated on only one side while others are marked on both sides. Major numbered graduations are in metres and tenths of metres. For example, the number 18 on the levelling rod is equal to 1.8 m. A pattern of squares, lines, spaces, and E shapes is used between the major numbers to indicate centimetres.

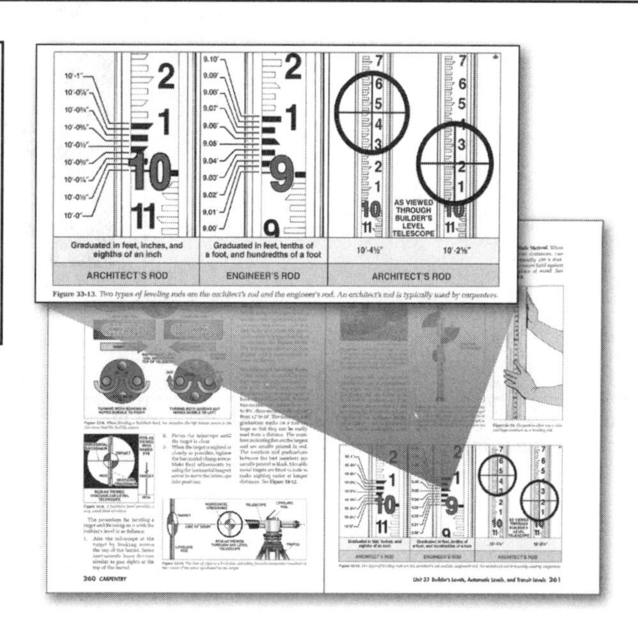

Metric Levelling Rod

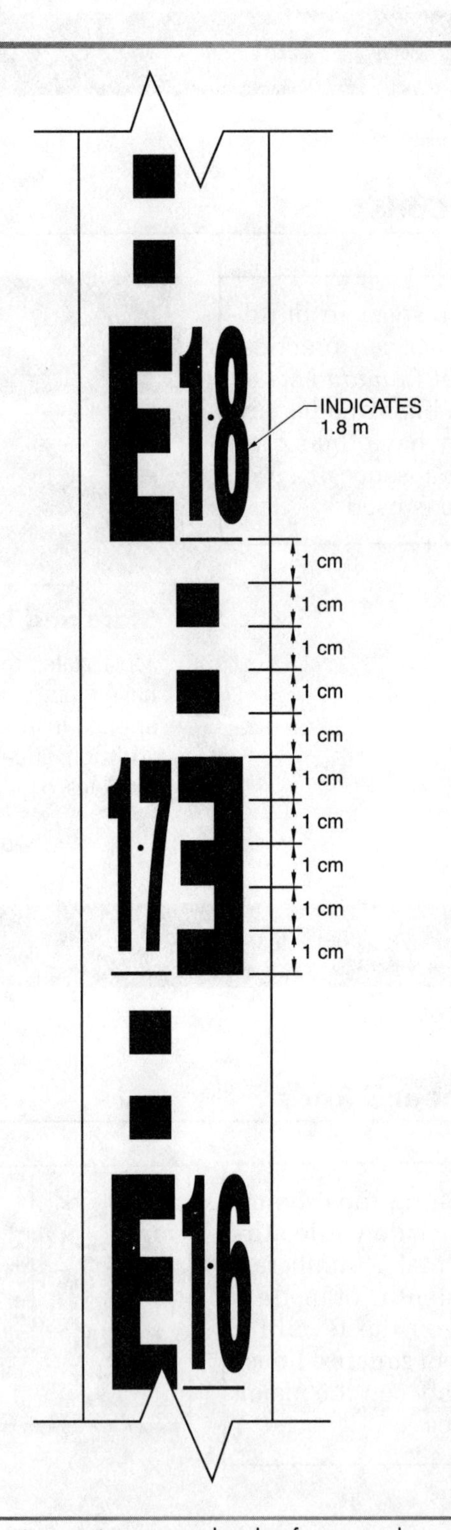

On a metric leveling rod, major numbered graduations are in metres and tenths of metres and a pattern of squares, lines, spaces, and E shapes indicates centimetres.

Foundation Sills

Page 294
Fastening Sill Plates with Anchor Bolts

The *National Building Code of Canada* requires a minimum bolt embedment of 100 mm (4″), spaced no more than 2 400 mm (7′10″) OC. There is no specified distance from the ends of a sill plate.

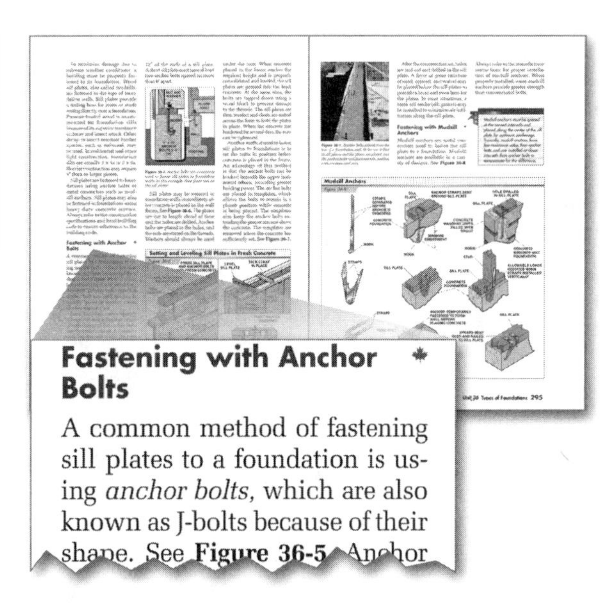

Fastening with Anchor Bolts

A common method of fastening sill plates to a foundation is using *anchor bolts,* which are also known as J-bolts because of their shape. See **Figure 36-5.** Anchor

Page 295
Fastening Sill Plates with Mudsill Anchors

Mudsill anchors are not specifically mentioned in the *National Building Code of Canada.* Embedment for anchor bolts is 100 mm (4″).

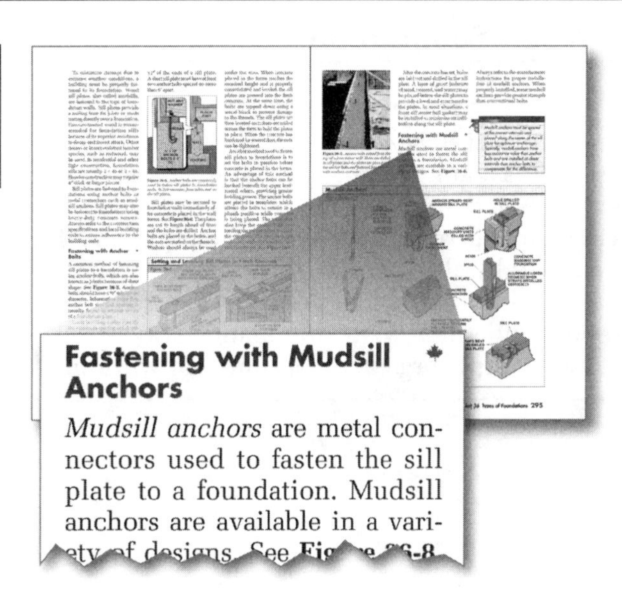

Fastening with Mudsill Anchors

Mudsill anchors are metal connectors used to fasten the sill plate to a foundation. Mudsill anchors are available in a variety of designs. See **Figure 36-8.**

Foundation Systems

Page 298
Foundations for Sloped Lots

The *National Building Code of Canada* requirements for stepped footings include a minimum horizontal distance between steps of 600 mm (23¾″) and a maximum vertical height per step of 600 mm (23¾″). The vertical footing must be at least 100 mm (4″) thick.

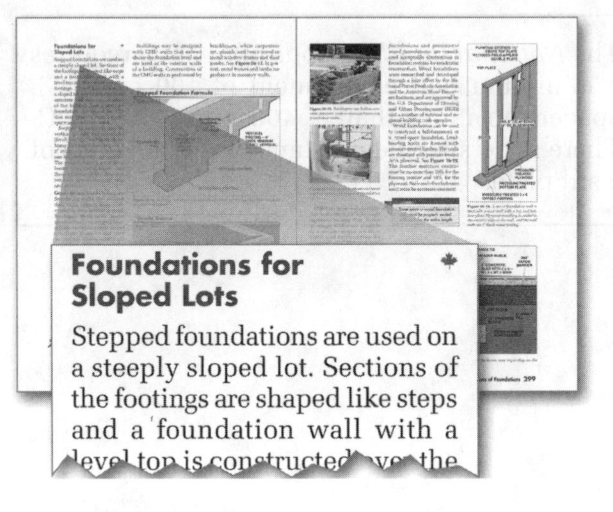

Wood Foundations

Page 299
Wood Foundations

The National Research Council Canada (NRC) publishes background information on preserved wood foundations. More information is available **http://irc.nrc-cnrc.gc.ca/pubs/cbd/cbd234_e.html.** Canadian Plywood (CANPLY) provides information on plywood for preserved wood foundations. More information is available at **www.canply.org/english/products/pwf.htm.**

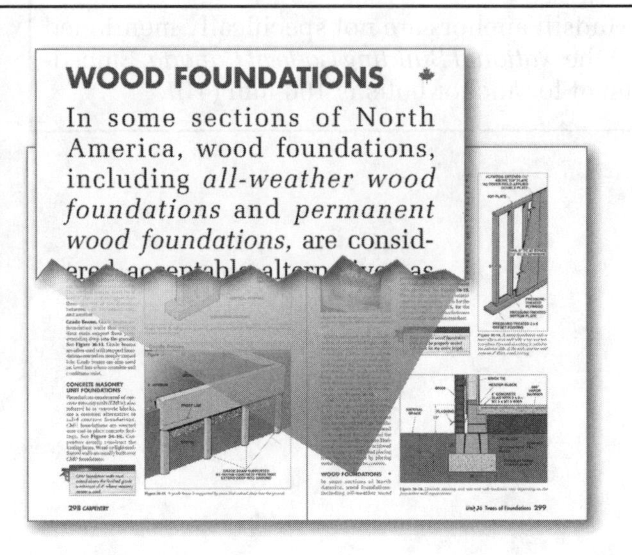

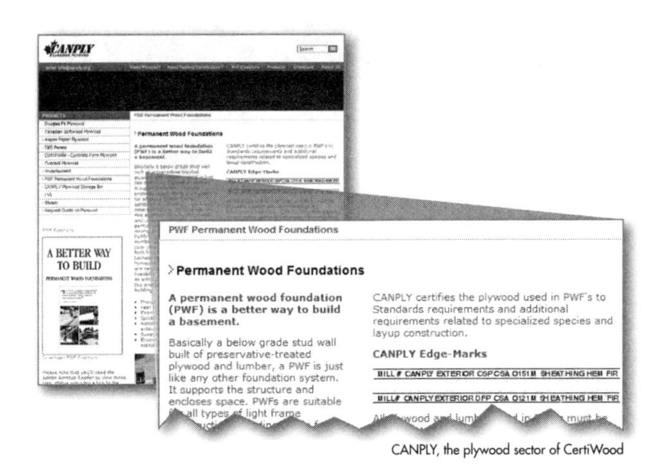

CANPLY, the plywood sector of CertiWood

See *Carpentry* CD-ROM Canadian Resources–National Research Council Canada (NRC)

See *Carpentry* CD-ROM Canadian Resources–CANPLY, the plywood sector of CertiWood

Reinforced Concrete

Page 304
Rebar

Metric rebar designation indicates the diameter of the rebar in millimetres, rounded to the nearest 5 mm.

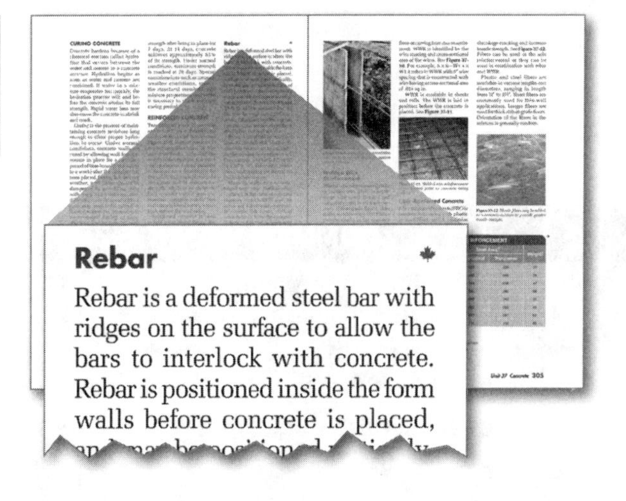

Rebar

Rebar is a deformed steel bar with ridges on the surface to allow the bars to interlock with concrete. Rebar is positioned inside the form walls before concrete is placed,

METRIC REBAR SIZES			
Size	**Mass***	**Diameter†**	**Cross-Section Area‡**
#10M	0.785	11.3	100
#15M	1.570	16.0	200
#20M	2.355	19.5	300
#25M	3.925	25.2	500
#30M	5.495	29.9	700
#35M	7.850	35.7	1000
#45M	11.775	43.7	1500
#55M	19.625	56.4	2500

* in kg/m
† in mm
‡ in mm²

Page 305
Common Stock Sizes of Welded Wire Reinforcement

In metric sizing for welded wire fabric, wire gauge or diameter is replaced by a cross-sectional area of the wire in square millimetres and the grid size is replaced with metric spacing dimensions (in millimetres).

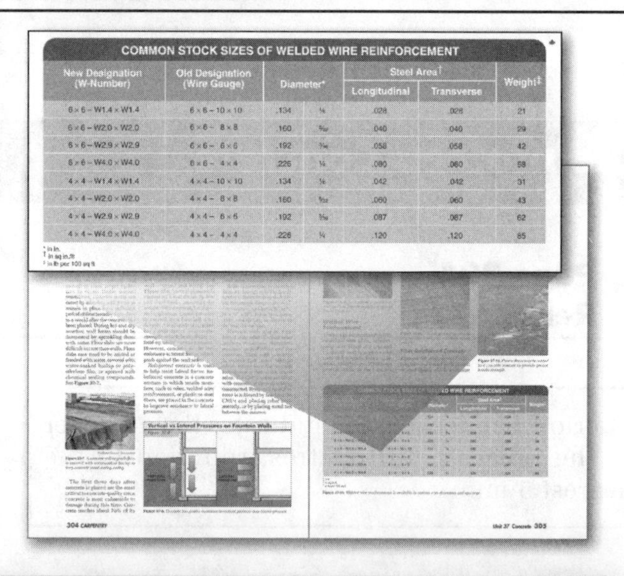

COMMON STOCK SIZES OF WELDED WIRE REINFORCEMENT						
New Designation (W-Number)	Old Designation (Wire Gauge)	Diameter*		Steel Area†		Weight‡
				Longitudinal	Transverse	
6 × 6 – W1.4 × W1.4	6 × 6 – 10 × 10	.134	1/8	.028	.028	21
6 × 6 – W2.0 × W2.0	6 × 6 – 8 × 8	.160	5/32	.040	.040	29
6 × 6 – W2.9 × W2.9	6 × 6 – 6 × 6	.192	3/16	.058	.058	42
6 × 6 – W4.0 × W4.0	6 × 6 – 4 × 4	.226	1/4	.080	.080	58
4 × 4 – W1.4 × W1.4	4 × 4 – 10 × 10	.134	1/8	.042	.042	31
4 × 4 – W2.0 × W2.0	4 × 4 – 8 × 8	.160	5/32	.060	.060	43
4 × 4 – W2.9 × W2.9	4 × 4 – 6 × 6	.192	3/16	.087	.087	62
4 × 4 – W4.0 × W4.0	4 × 4 – 4 × 4	.226	1/4	.120	.120	85

* in in.
† in sq in./ft
‡ in lb per 100 sq ft

COMMON STOCK SIZES OF WELDED WIRE FABRIC		
New Designation (W-Number) (Wire Spacing and Cross-Section Area)	**Old Designation (Wire Spacing and Gauge)**	**Metric Designation (Wire Spacing and Cross-Section Area)**
6 × 6 – W1.4 × W1.4	6 × 6 – 10 × 10	152 × 152 – MW9.1 × 9.1
6 × 6 – W2.0 × W2.0	6 × 6 – 6 × 6	152 × 152 – MW18.7 × 18.7
6 × 6 – W2.9 × W2.9	6 × 6 – 4 × 6	152 × 152 – MW25.8 × 18.7
6 × 6 – W4.0 × W4.0	6 × 6 – 4 × 4	152 × 152 – MW25.8 × 25.8
4 × 4 – W1.4 × W1.4	4 × 4 – 10 × 10	102 × 102 – MW9.1 × 9.1
4 × 4 – W2.0 × W2.0	4 × 4 – 8 × 8	102 × 102 – MW13.3 × 13.3
4 × 4 – W2.9 × W2.9	4 × 4 – 6 × 6	102 × 102 – MW18.7 × 18.7
4 × 4 – W4.0 × W4.0	4 × 4 – 4 × 4	102 × 102 – MW25.8 × 25.8

Compiled with information provided by Canada Plan Service

Page 339
Termites

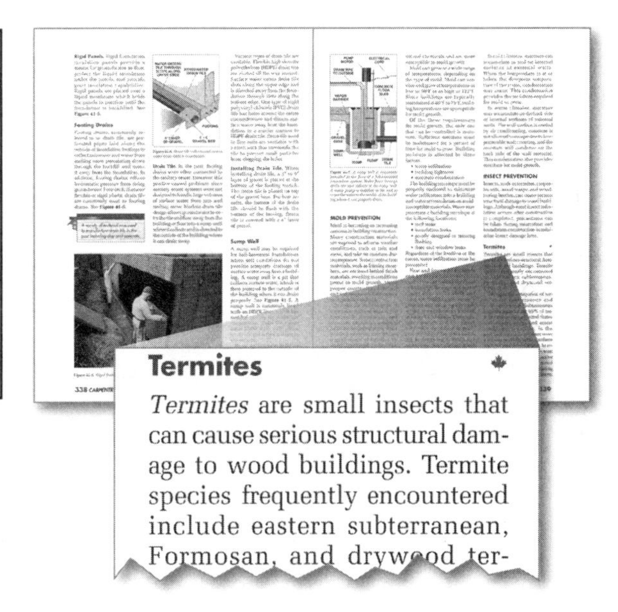

Termites

Termites are small insects that can cause serious structural damage to wood buildings. Termite species frequently encountered include eastern subterranean, Formosan, and drywood ter-

In most areas of Canada, termite infestations are not common, and protective measures are not required except in specified local areas. The areas of Canada in which infestations occur are identified on the map. When required, protective measures are normally enforced only by affected municipalities. The National Research Council Canada identifies areas where termites are known to occur in *Seminars on the Technical Changes in the 2005 Construction Codes.*

Termite and Decay Protection

National Research Council Canada

Posts and Beams

Page 346
Wood Beams

Building codes in Canada use similar tables to establish beam spans, but wood species and lumber grading conform to Canadian practice. Consult the *National Building Code of Canada*, the appropriate provincial building code, or an approved book of span tables such as the *Canadian Span Book*, published by the Canadian Wood Council.

Page 348
Determining Allowable Beam Sizes

Beam sizes for Canadian buildings should be taken from an approved span book or the applicable Canadian building code rather than those shown in Figure 42-8.

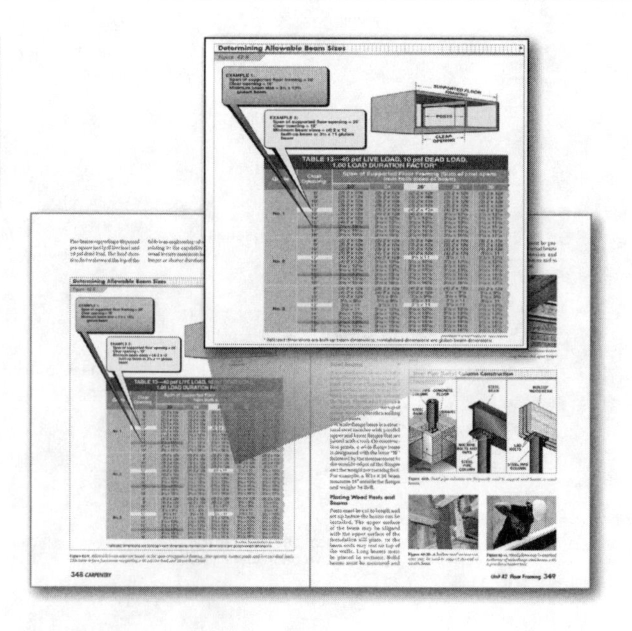

Page 349
Steel Beams

Wide-flange steel beams used in Canadian construction are designated by metric size and mass (weight). For example, a W200 × 27 beam has a nominal depth across flanges of 200 mm and weighs 27 kg/m. To convert from one system to the other, it is necessary to convert the depth across the flanges and the mass.

Steel Beams

A standard steel beam, called a *wide-flange beam,* is commonly used with wood framing. Wood joists either rest on top of the beam or butt against the sides of

Floor Joists

Page 350
Allowable Joist Spans

To determine allowable spans for joists in Canadian residential construction, refer to the *National Building Code of Canada*, the applicable provincial building code, or the CWC span book.

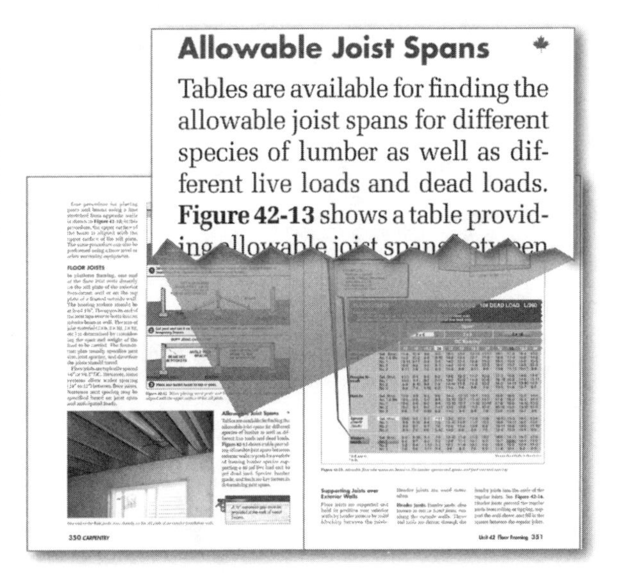

Allowable Joist Spans

Tables are available for finding the allowable joist spans for different species of lumber as well as different live loads and dead loads. **Figure 42-13** shows a table provid-

Subfloor

Page 364
Subfloor

Canadian Plywood (CANPLY) provides information on tongue and groove plywood sheathing. More information is available at **www.canply.org/english/products/easy_tg/easytgfloor.htm.** The Canadian Wood Council publishes information on panel marks and OSB waterboard in *Wood Reference Handbook*.

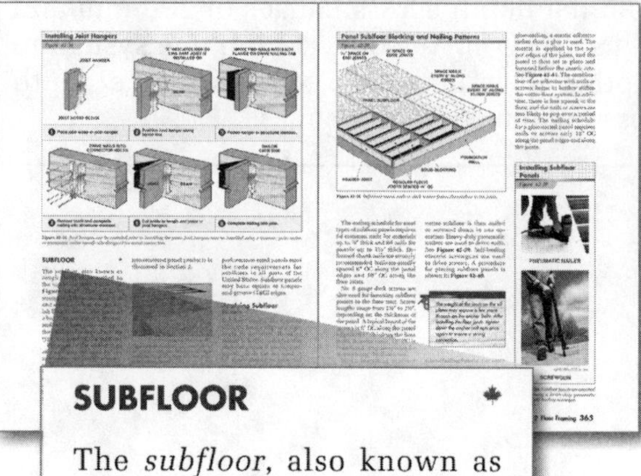

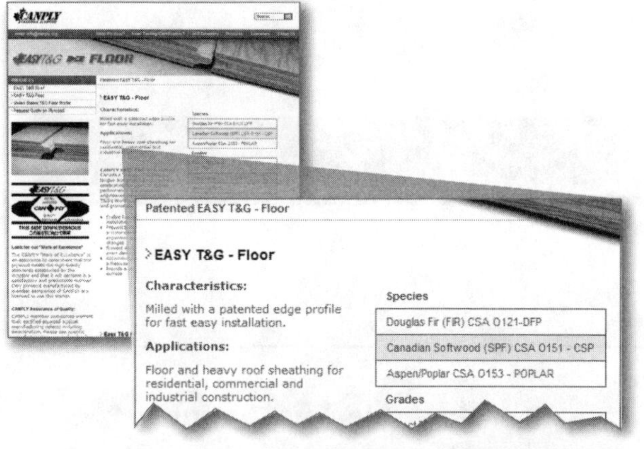

CANPLY, the plywood sector of CertiWood

See Carpentry CD-ROM Canadian Resources–CANPLY, the plywood sector of CertiWood

Panel Marks for OSB and Waterboard

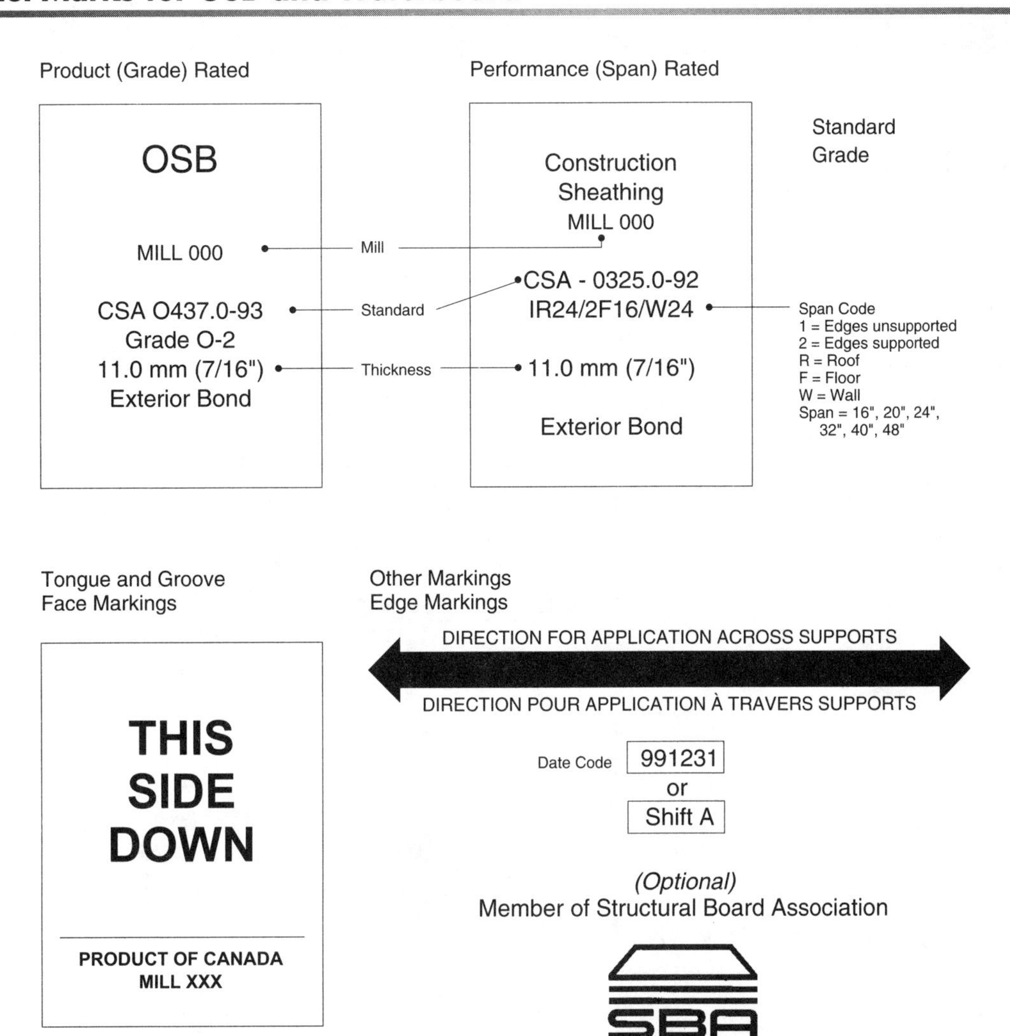

Excerpt from *Wood Reference Handbook,* ©1991 by Canadian Wood Council

Floor Underlayment

Page 366 Floor Underlayment

Canadian Plywood (CANPLY) provides information on unsanded plywood panels, which are functionally equivalent to Sturd-I-Floor™ panels. More information is available at **www.canply.org/english/products/csp/csp_unsanded_stf.htm.**

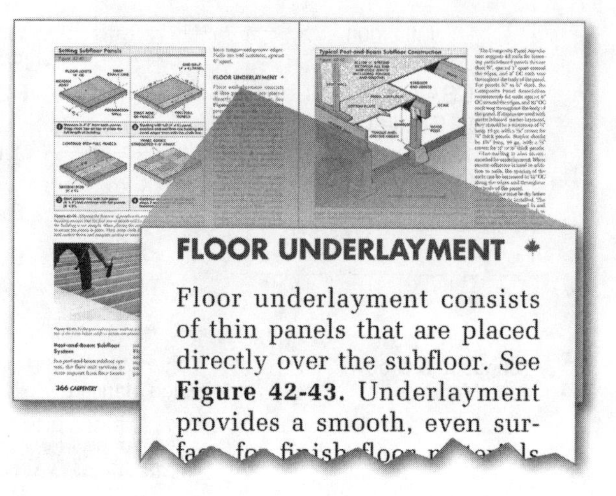

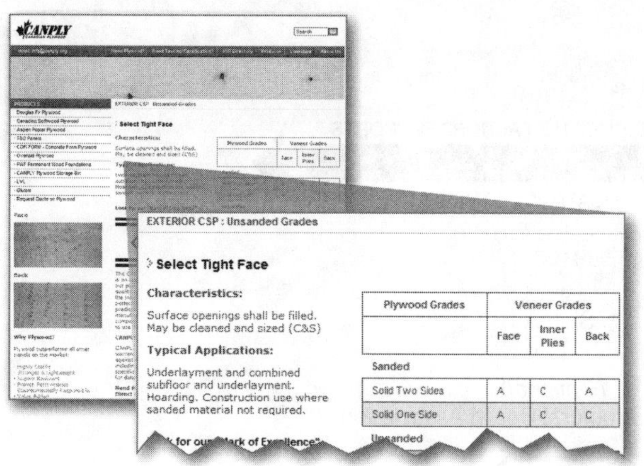

CANPLY, the plywood sector of CertiWood

See Carpentry CD-ROM Canadian Resources–CANPLY, the plywood sector of CertiWood

Page 366 Fastening Methods for Underlayment

Joist spans for Canadian construction should be determined from Canadian codes.

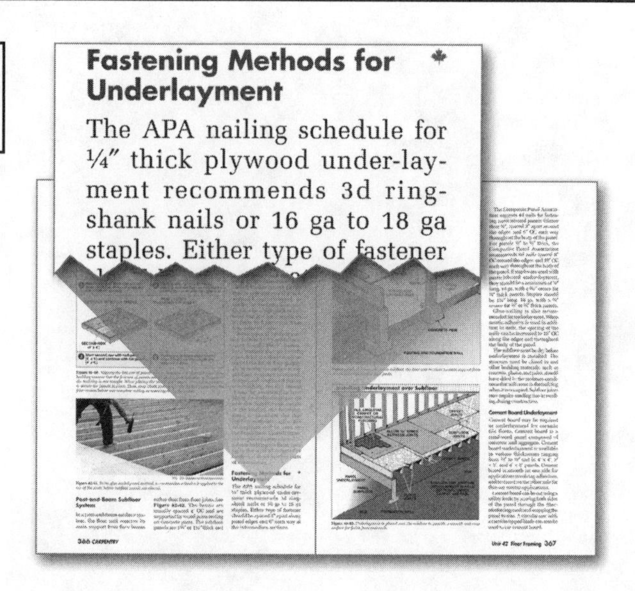

Wood I-Joists

Page 368
Wood I-Joists

The Canadian Construction Materials Centre (CCMC) provides a searchable registry of product evaluations from the National Research Council. More information is available at **http://irc.nrc-cnrc.gc.ca/ccmc/regprodeval_e.html.**

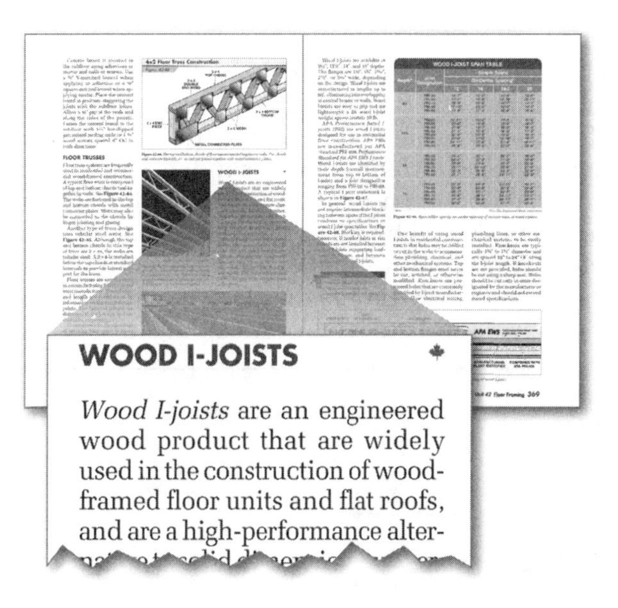

WOOD I-JOISTS

Wood I-joists are an engineered wood product that are widely used in the construction of wood-framed floor units and flat roofs, and are a high-performance alter-

See Carpentry CD-ROM Canadian Resources–Canadian Construction Materials Centre (CCMC)

Sheathing Exterior Walls

Page 402 Wall Sheathing

Canadian Plywood (CANPLY) provides information on poplar and aspen sheathing. More information is available at **www.canply.org/english/products/pop/pop_unsanded_shg.htm.** Canadian code requirements for fastening wall sheathing should be adhered to. The spacing of fasteners (excluding staples) is similar to U.S. requirements, but the length of fasteners required for particular thicknesses of sheathing differs.

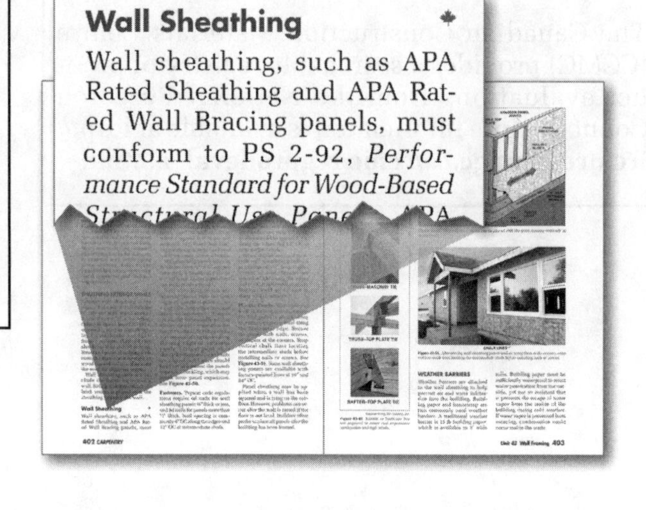

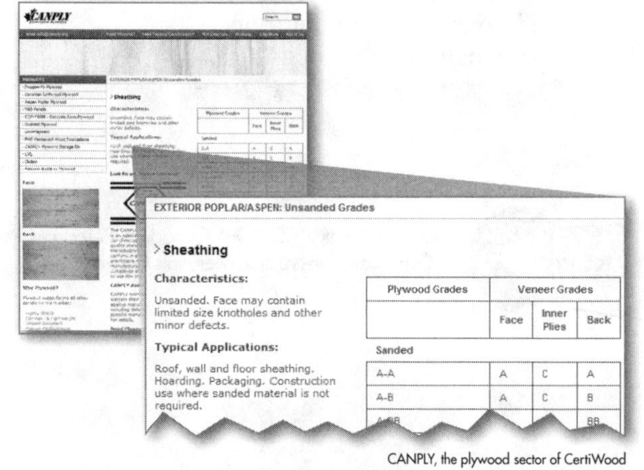

CANPLY, the plywood sector of CertiWood

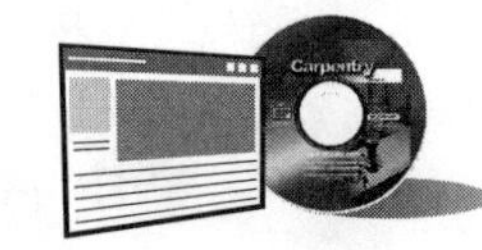

See Carpentry CD-ROM Canadian Resources–CANPLY, the plywood sector of CertiWood

Shear Walls

Page 407
Constructing Shear Walls

Requirements for construction of shear walls for Canadian buildings are not specified in building codes. Specifications for these and other specialized building components must be determined by a qualified designer using engineering data from sources such as the Canadian Wood Councils's *Wood Reference Handbook.*

Constructing Shear Walls

As shown in **Figure 43-61**, shear wall construction is similar to conventional wood-framed walls, but there are several important differences, including the following:

Ceiling Joists

Page 411
Ceiling Joist Spans

Ceiling joist spans for Canadian residences must be determined from tables found at the back of Part 9 of the National Building Code of Canada, provincial building codes, or an approved span book, rather than those shown in Figure 44-2.

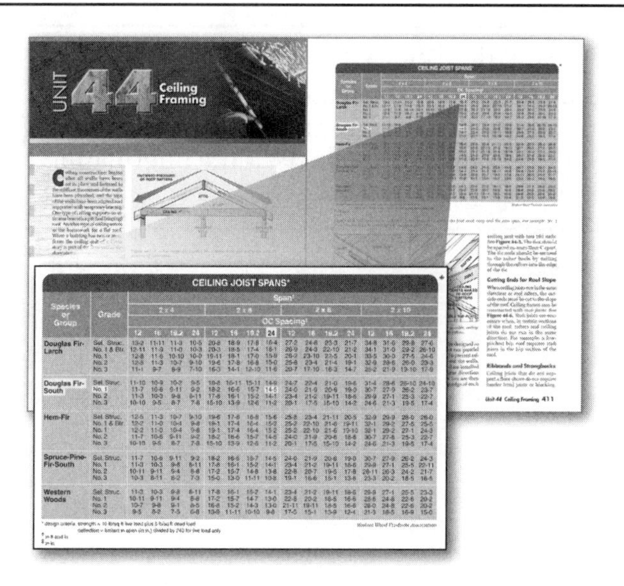

Constructing Flat Roof Ceilings

Page 416
Constructing Flat Roof Ceilings

Canadian codes typically require higher load ratings for flat roof ceilings. The low end of the range is 1.0 kPa, equivalent to 21 lb/sq ft, with requirements as high as 3.0 kPa (63 lb/sq ft).

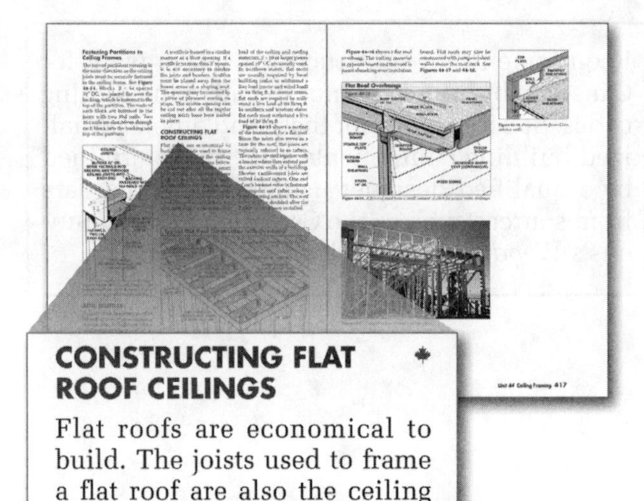

CONSTRUCTING FLAT ROOF CEILINGS

Flat roofs are economical to build. The joists used to frame a flat roof are also the ceiling joists for the living area below.

Light-Gauge Steel Framing Members

Page 419
Shapes and Dimensions

The Canadian Sheet Steel Building Institute (CSSBI) is a trade association providing information and resources for manufacturers and users of sheet steel building products. More information is available at **www.cssbi.ca.**

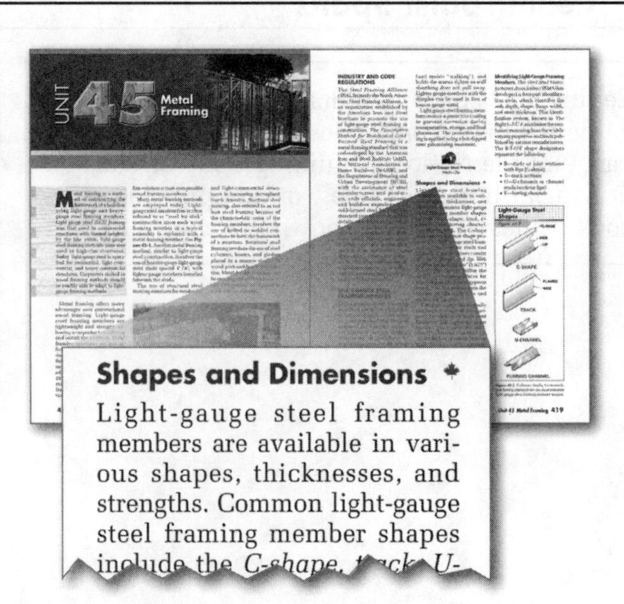

Shapes and Dimensions

Light-gauge steel framing members are available in various shapes, thicknesses, and strengths. Common light-gauge steel framing member shapes

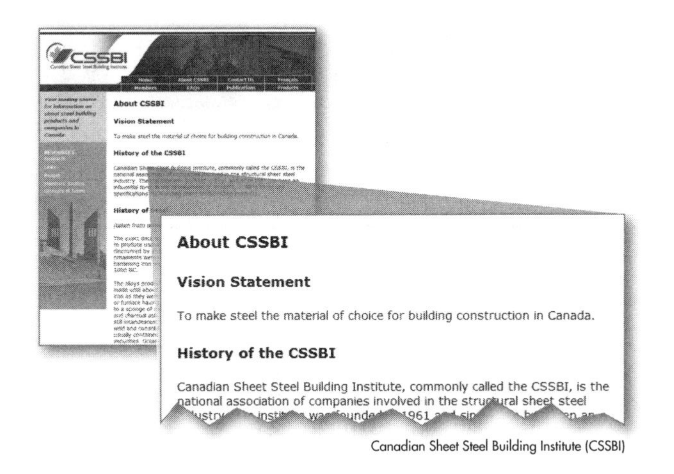

Canadian Sheet Steel Building Institute (CSSBI)

See Carpentry CD-ROM Canadian Resources–Canadian Sheet Steel Building Institute (CSSBI)

Structural Factors in Roof Design

Page 444
Allowable Rafter Spans

Rafter spans for Canadian construction must be determined from the *National Building Code of Canada,* provincial codes, or an approved span book. Live loads on roofs vary across Canada and include both snow and rain loads. To determine which load factor to use in the rafter tables, the designer must first locate climatic data for the location and then apply a formula found in the *National Building Code of Canada.*

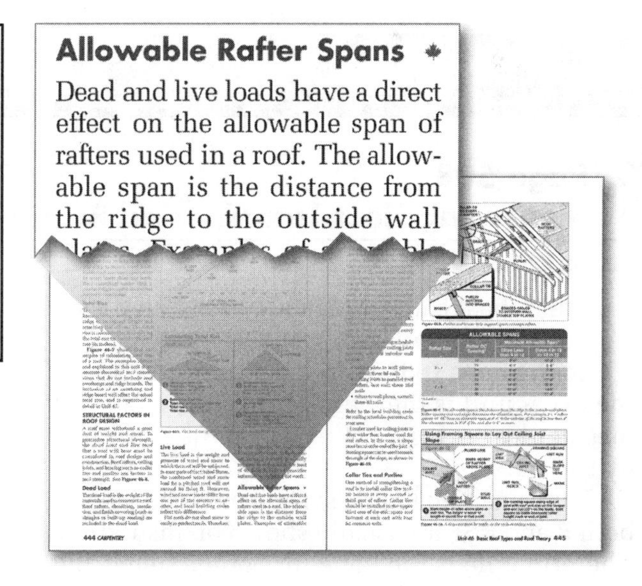

Installing Roof Trusses

Page 501
Bracing Trusses

The Canadian Wood Truss Association (CWTA) is a national trade association for wood truss manufacturers. More information is available at **www.cwta.net.**

Bracing Trusses

Proper temporary bracing while placing trusses is essential to prevent collapse of the trusses, with possible damage to the trusses and worker injury. Inadequate

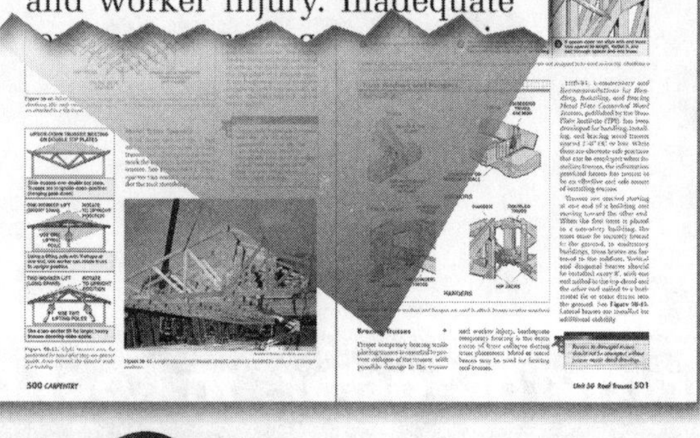

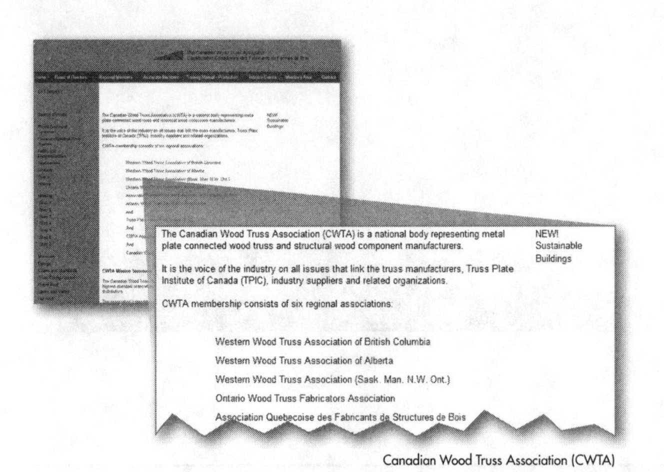

Canadian Wood Truss Association (CWTA)

See Carpentry CD-ROM Canadian Resources–Canadian Wood Truss Association (CWTA)

Heat Transfer

Page 508
Measuring Heat Transfer and Resistance

Canadian heat calculations are based on a metric unit, the joule (J). One Btu is equivalent to approximately 1052 J. To keep numbers small, the usual unit is the megajoule (MJ). To convert megajoules to British thermal units, multiply by 947.82.

The Canadian metric system uses RSI values; both R and RSI are usually shown on insulation. The principle is the same for both systems, but resistance to heat transfer is calculated using metric units. To convert RSI to R, multiply by 5.6785.

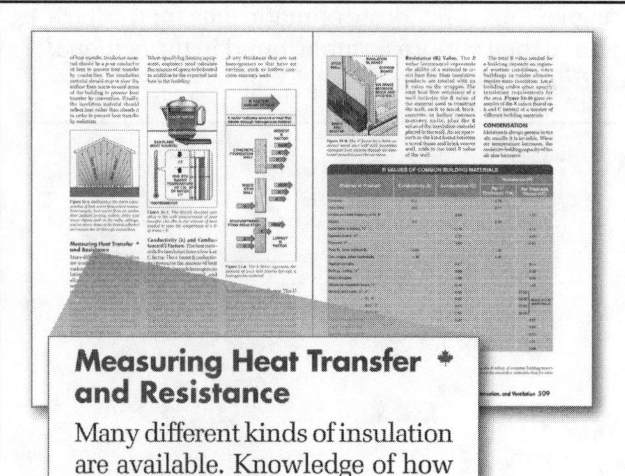

Measuring Heat Transfer and Resistance

Many different kinds of insulation are available. Knowledge of how various materials used for insulation resist the passage of heat

Thermal Insulation

Page 515
Recommended Insulation R Values

Insulation requirements in Canada are based on degree days and tied to long-term average daily temperatures observed over the heating season. The difference between 18°C and the actual average temperature below 18°C is recorded; this total per year is averaged over many years and used to produce tables and maps. This information is used to determine insulation requirements for various areas in Canada. The National Research Council Canada (NRC) publishes information related to heating degree days for various geographical regions across Canada. More information is available at **http://atlas.nrcan.gc.ca/site/english/maps/archives/5thedition/environment/climate/mcr4033.**

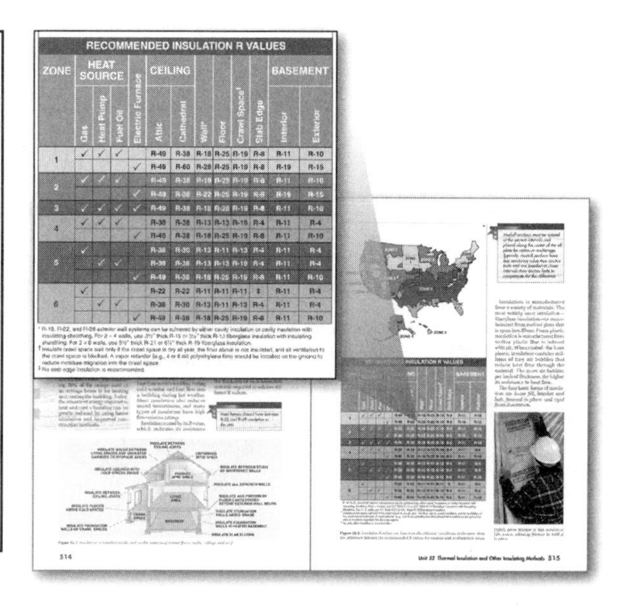

RECOMMENDED INSULATION R VALUES

ZONE	HEAT SOURCE: Gas	HEAT SOURCE: Heat Pump	HEAT SOURCE: Fuel Oil	Electric Furnace	CEILING: Attic	CEILING: Cathedral	Wall*	Floor	Crawl Space†	Slab Edge	BASEMENT: Interior	BASEMENT: Exterior
1	✓	✓	✓		R-49	R-38	R-18	R-25	R-19	R-8	R-11	R-10
				✓	R-49	R-60	R-28	R-25	R-19	R-8	R-19	R-15
2	✓	✓	✓		R-49	R-38	R-18	R-25	R-19	R-8	R-11	R-10
				✓	R-49	R-38	R-22	R-25	R-19	R-8	R-19	R-15
3	✓	✓	✓	✓	R-49	R-38	R-18	R-25	R-19	R-8	R-11	R-10
4	✓	✓	✓		R-38	R-38	R-13	R-13	R-19	R-4	R-11	R-4
				✓	R-49	R-38	R-18	R-25	R-19	R-8	R-11	R-10
5	✓				R-38	R-30	R-13	R-11	R-13	R-4	R-11	R-4
		✓	✓		R-38	R-38	R-13	R-13	R-19	R-4	R-11	R-4
				✓	R-49	R-38	R-18	R-25	R-19	R-8	R-11	R-10
6	✓				R-22	R-22	R-11	R-11	R-11	‡	R-11	R-4
		✓	✓		R-38	R-30	R-13	R-11	R-13	R-4	R-11	R-4
				✓	R-49	R-38	R-18	R-25	R-19	R-8	R-11	R-10

* R-18, R-22, and R-28 exterior wall systems can be achieved by either cavity insulation or cavity insulation with insulating sheathing. For 2 × 4 walls, use 3½" thick R-15 or 3½" thick R-13 fiberglass insulation with insulating sheathing. For 2 × 6 walls, use 5½" thick R-21 or 6¼" thick R-19 fiberglass insulation.
† Insulate crawl space wall only if the crawl space is dry all year, the floor above is not insulated, and all ventilation to the crawl space is blocked. A vapor retarder (e.g., 4 or 6 mil polyethylene film) should be installed on the ground to reduce moisture migration into the crawl space.
‡ No slab edge insulation is recommended.

Annual Sum of Heating Degree Days

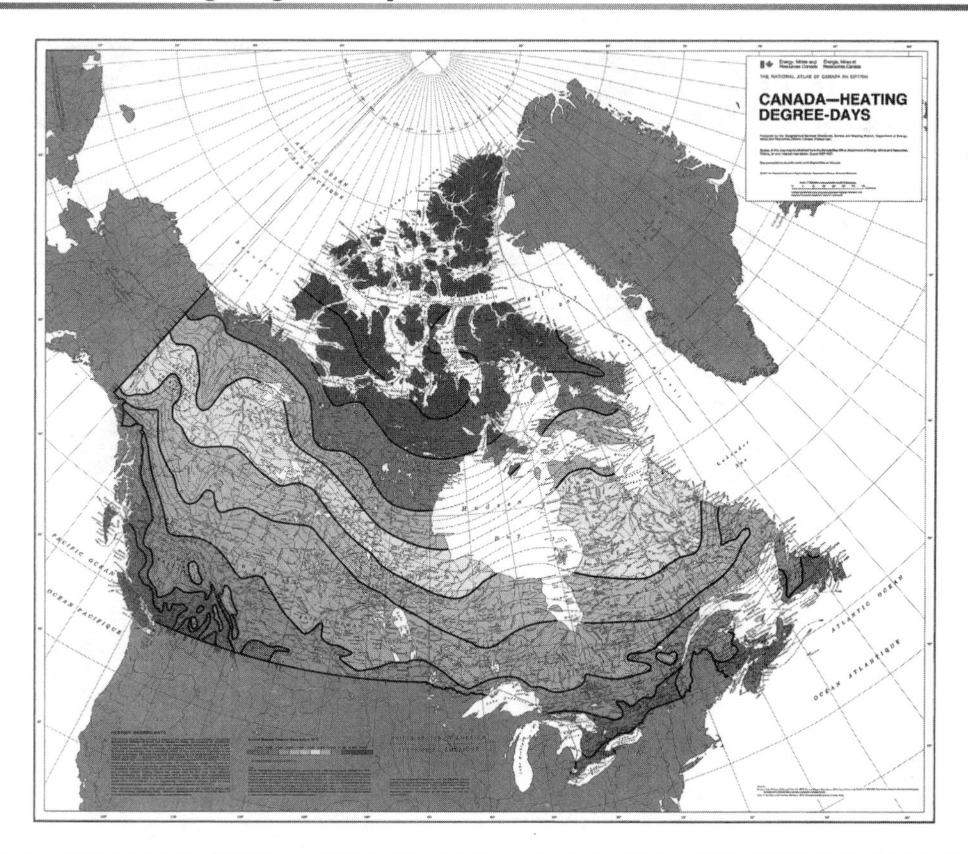

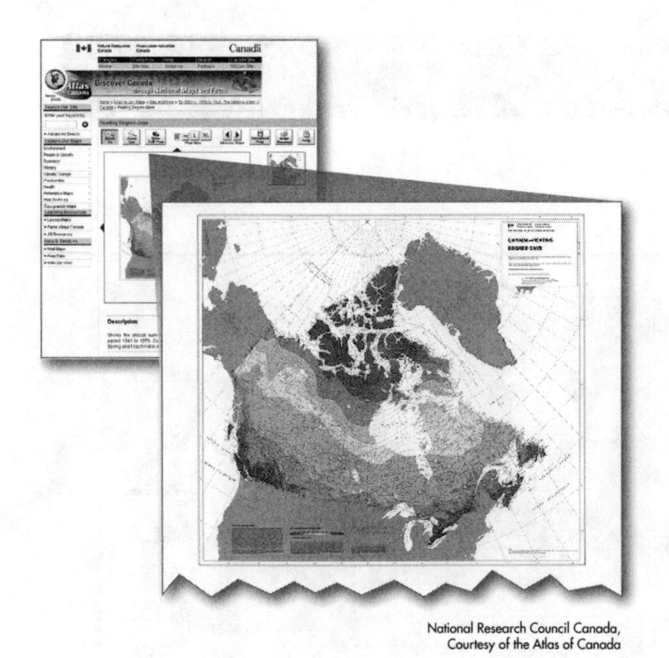

National Research Council Canada,
Courtesy of the Atlas of Canada

See Carpentry CD-ROM Canadian Resources–
National Research Council Canada (NRC)

Covering Roofs

Page 550
Asphalt Strip Shingles

Asphalt shingles produced in Canada measure 1 metre by ⅓ metre, or 39⅜″ by 13¼″, rather than the sizes shown in Figure 55-13.

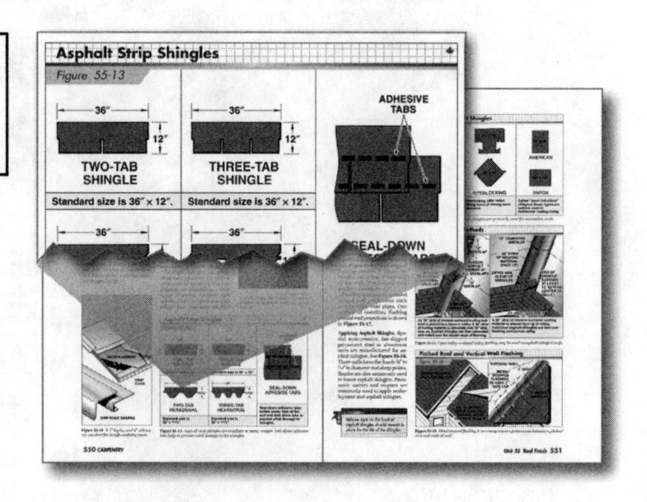

Page 553
Wood Shingles and Shakes

Canadian building codes restrict exposure for shakes to a maximum of 240 mm or 9½″.

Spacing between shakes is limited to 6 mm (¼″) to 9 mm (⅜″) by Canadian codes.

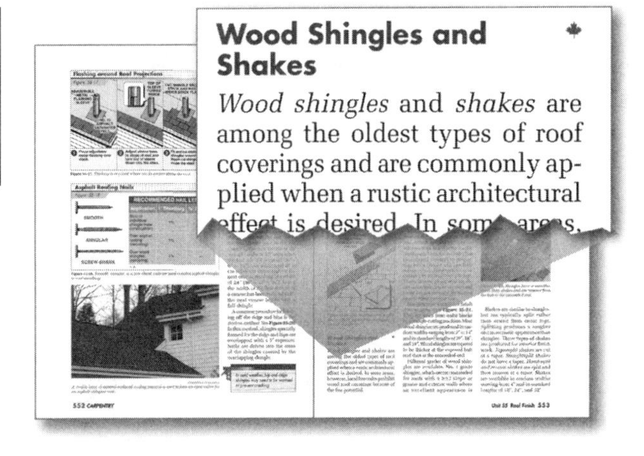

Wood Shingles and Shakes

Wood shingles and *shakes* are among the oldest types of roof coverings and are commonly applied when a rustic architectural effect is desired. In some areas,

Page 554
Spaced Sheathing under Wood Shakes or Shingles

Figure 55-22 shows spacing between shakes as ⅜″ to ½″; the *National Building Code of Canada* specifies spacing of 6 mm (¼″) to 9 mm (⅜″).

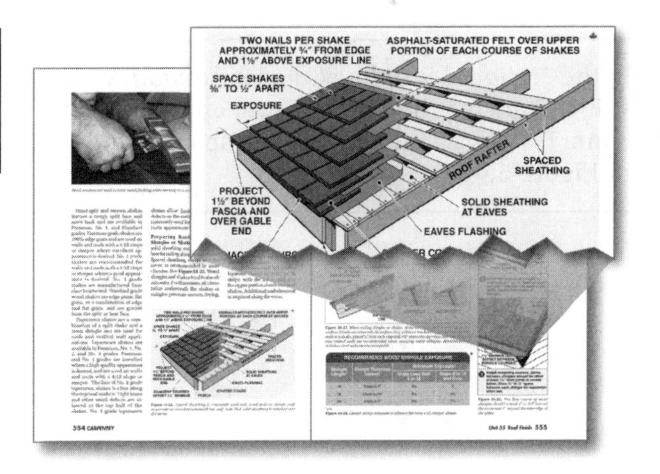

Page 555
Recommended Wood Shingle Exposure

Consult Canadian building codes for allowable exposures, as the requirements differ from those shown in Figure 55-24.

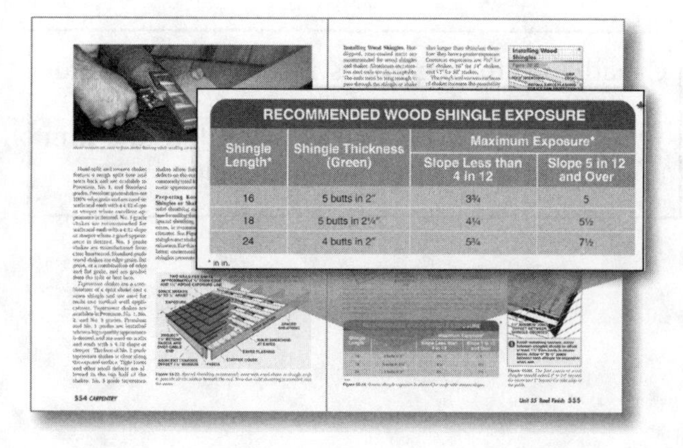

RECOMMENDED WOOD SHINGLE EXPOSURE

Shingle Length*	Shingle Thickness (Green)	Maximum Exposure*: Slope Less than 4 in 12	Slope 5 in 12 and Over
16	5 butts in 2"	3¾	5
18	5 butts in 2¼"	4¼	5½
24	4 butts in 2"	5¾	7½

* in in.

Page 555
Installing Wood Shingles

In the *National Building Code of Canada,* wood shingles are required to be spaced approximately 6 mm (¼″) apart, rather than the spacing shown in Figure 55-25.

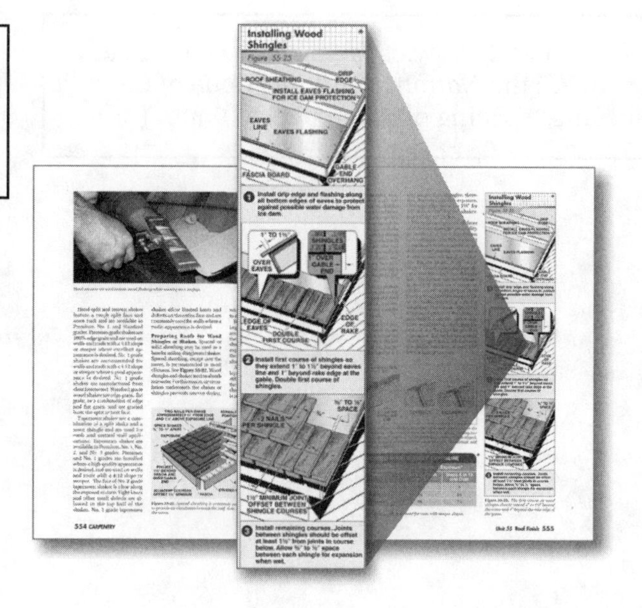

Window Units

Page 563
Window Selection

There is no Canadian equivalent to the National Fenestration Rating Council. Specifications and standards for windows, doors, and many other building components are developed in Canada by the Canadian Standards Association (CSA) and the Canadian General Standards Board (CGSB). In addition, standards established by the American Society for Testing and Materials (ASTM) are used in Canada. The standard governing window is CSA A441.2.

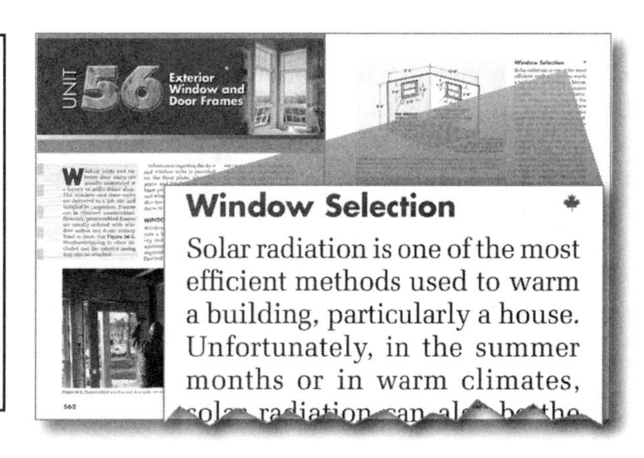

Page 564
CSA Approved Certification Marks

CSA International publications and web site show approved certification marks in use in Canada and the United States. More information is available at **www.csa-international.org/certification_marks/marks_for_canada.**

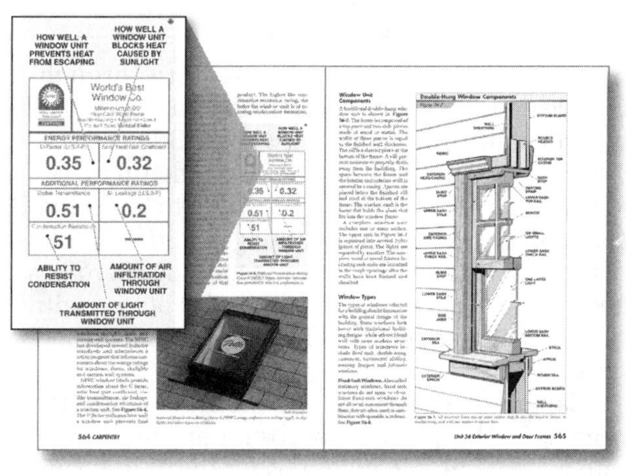

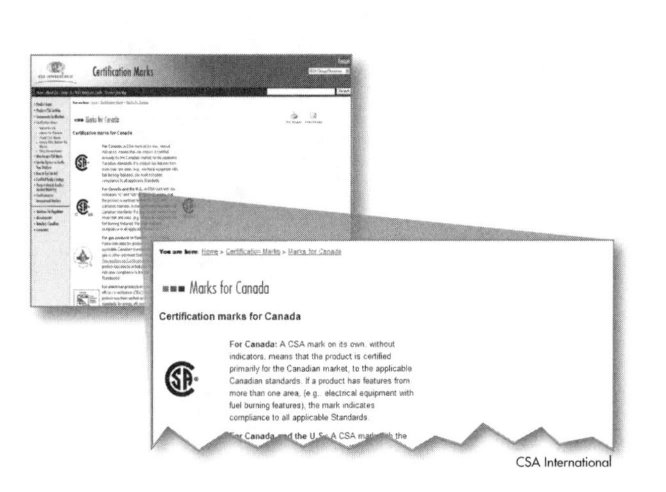

CSA International

See Carpentry CD-ROM Canadian Resources – CSA International

Wood Siding

Page 589
Panel Siding

Canadian requirements for fastening of panel siding, as well as spacing between panels, may differ from those stated in this unit. Appropriate Canadian sources should be consulted.

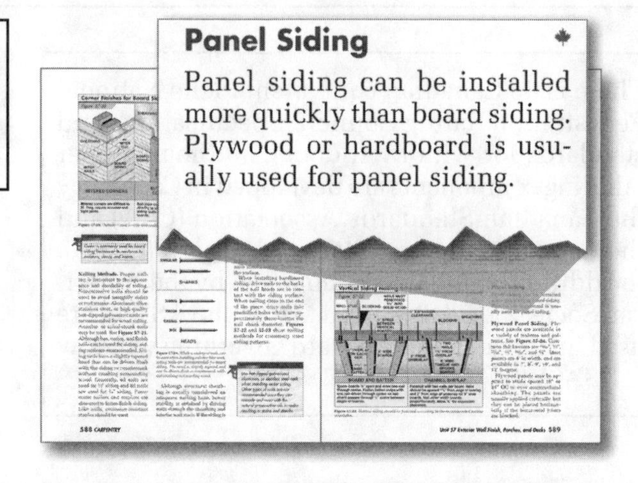

Decks

Page 604
Constructing Decks

Some Canadian jurisdictions require through-bolting deck ledgers to joists in the building. This is particularly true for homes with OSB rim boards, where a backing board may be needed between joists. It may also not be permissible to bolt a ledger to brick veneer. If a sheathing material such as gypsum board, fibreboard, or insulating board has been used, it must be removed and replaced with a rigid material such as OSB or plywood. There must not be a gap behind a ledger board.

Canadian codes do not include sandwich beams, as the plies of the beam are not nailed into a single unit. Guidelines for built-up beams should be followed. In addition, supporting beam components by nailing or bolting to the sides of a post, without the use of ledgers or hangers, may not be acceptable. The *National Building Code of Canada* requires exposed decks in a given area to carry the same snow and rain loads as roof structures. In most cases, this will require decks to be built to the same standard as the main living areas of a home.

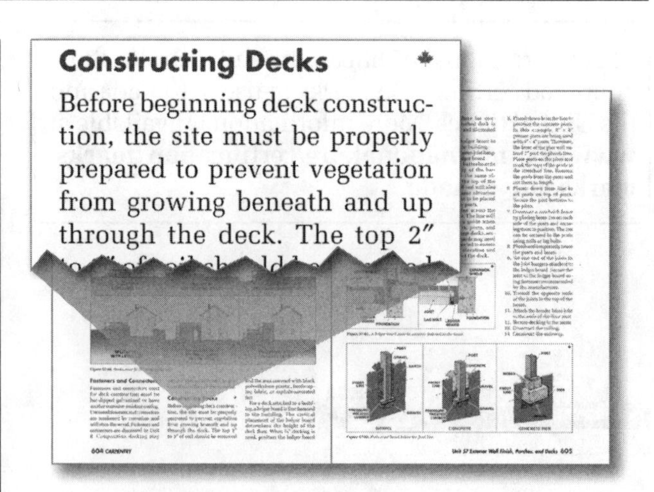

Page 605
Fastening Ledgers to Buildings

Methods illustrated in Figure 57-61 may not be acceptable in all Canadian jurisdictions. Check local codes for acceptable ledger attachments.

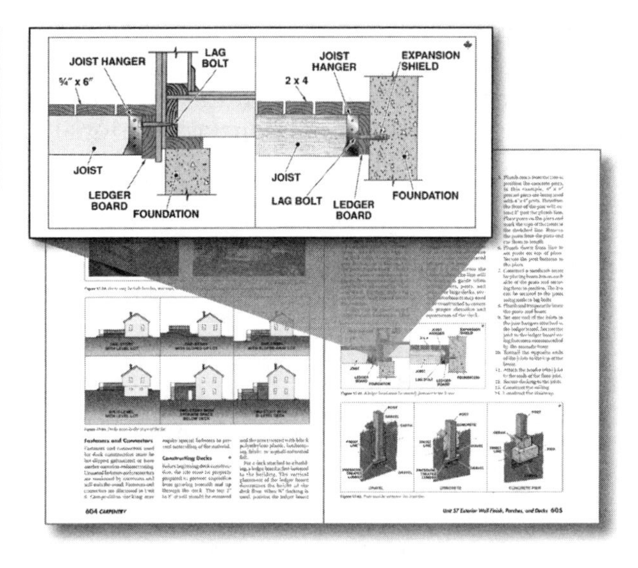

Page 605
Setting Posts below Frost Line

Post footings illustrated in Figure 57-62 may not be accepted in some Canadian municipalities where the deck height is greater than 600 mm (24"). Concrete footings and piers may be required, with the pier cast monolithically with the footing, or attached to it with rebar. The pier may be required to be tapered, or flared at the bottom, to prevent ad-freezing and lifting caused by frost.

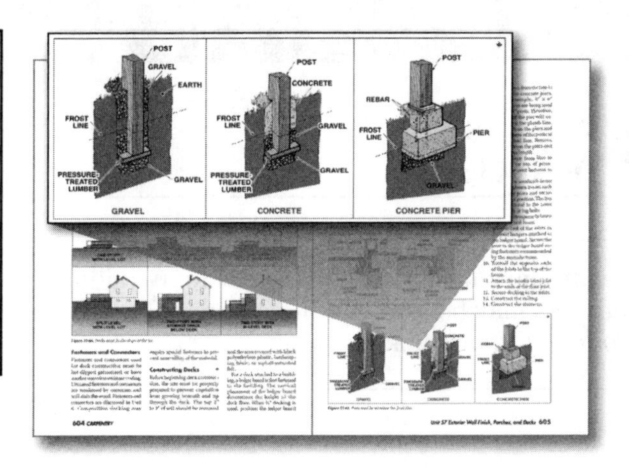

Page 606
Constructing Decks

Some structural details illustrated in Figure 57-63 may not be acceptable in Canadian construction. Consult the appropriate national, provincial, and local codes.

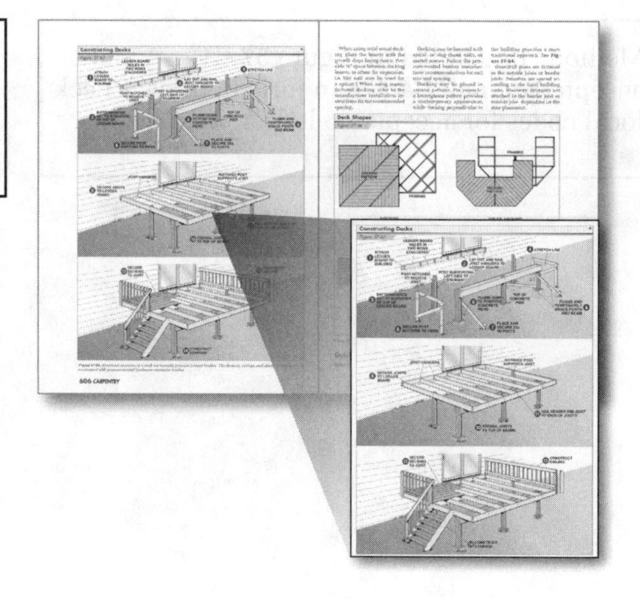

Other Door Hardware

Page 652
Door Closers

Headroom clearance requirements for Canadian builders may be found in the *National Building Code of Canada,* Part 3.

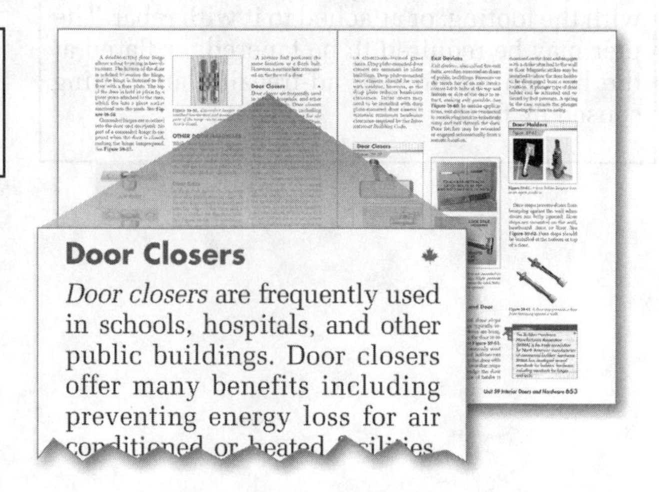
Door Closers

Door closers are frequently used in schools, hospitals, and other public buildings. Door closers offer many benefits including preventing energy loss for air conditioned or heated facilities

Page 653
Canadian Hardware and Housewares Manufacturers Association

The Canadian Hardware and Housewares Manufacturers Association (CHHMA) is the trade association that represents manufacturers of building hardware, as well as a wide range of other products.

Stairway Components

Page 687
Stringer Layout

The *National Building Code of Canada* requirements for stair rise and run are somewhat different. Maximum riser height is 200 mm (7⅞″), with a minimum tread depth of 235 mm (9¼″) for residential stairs. Minimum riser height is 125 mm (4⅞″).

Stringer Layout

Stringer layout includes marking off treads and risers. All risers should be the same height and all treads should be the same depth to ensure safe and smooth movement from one level to the next.

Unit 64 Stairway Construction

Safety is a major concern in the design of a stairway. A high percentage of accidents occurring at home take place on stairways. Most building codes include detailed and strict regulations for stairway construction.

Refer to your local building code for information regarding specific stairway requirements for your geographic area.

CONSTRUCTING INTERIOR STAIRWAYS

Stairways must be constructed in accordance with the local building code. Stairway requirements are different for residential and commercial construction. The *National Building Code of Canada* provides stairway requirements for dwellings and public and commercial buildings, or buidings with public areas.

Residential Stairway Requirements

Building codes address several residential stairway requirements including stairway width, headroom, treads and risers, profiles, landings, and handrails.

Stairway Width. Stairways for residential structures should not be less than 860 mm (34″) wide at all points between the top of a handrail and required headroom height. See **Figure 64-1.** Handrails should not project more than 100 mm (4″) on either side of a stairway.

Headroom Requirements. *Headroom* is the minimum vertical clearance between tread nosings and the ceiling above. The minimum headroom for residential stairways should not be less than 1 950 mm (6′-5″) measured vertically from a line connecting the edges of the nosings. See **Figure 64-2.** The minimum headroom must be maintained between parallel flights of stairs. See **Figure 64-3.**

MINIMUM CLEAR WIDTH ABOVE HANDRAILS = 860 mm (34″)

Figure 64-1. *The minimum stairway width between wall faces is 860 mm (34″).*

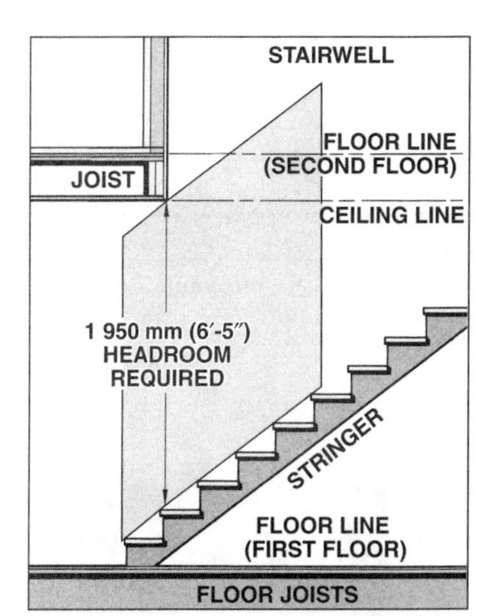

Figure 64-2. *Headroom is the minimum vertical clearance required and is measured from a line connecting the nosings on the stairway to any part of the ceiling above the stairway.*

Riser Heights and Tread Depths. The maximum riser height for residential stairways is 200 mm ($7\frac{7}{8}$″) measured vertically between the nosings of adjacent treads. See **Figure 64-4.** In addition, the variation in riser heights between the largest and smallest riser height should not be more than 6 mm (1/4″). The minimum tread depth for residential stairways is 235 mm (9 1/4″) with the tread depth measured horizontally from the nosings of adjacent treads. The variation in tread depth between the largest and smallest tread depth should not be more than 6 mm (1/4″).

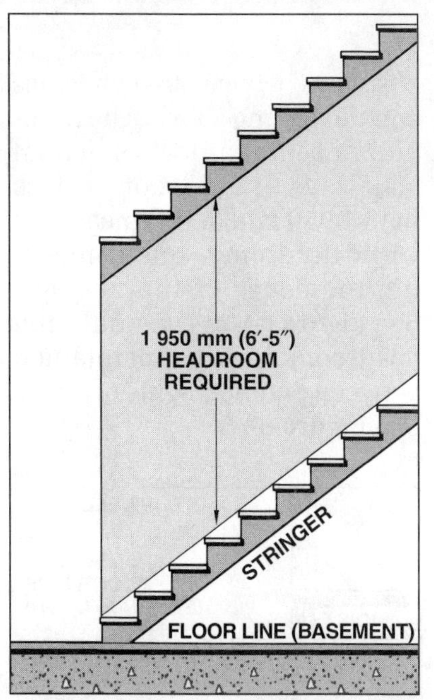

Figure 64-3. *The minimum headroom must be maintained between parallel flights of stairs.*

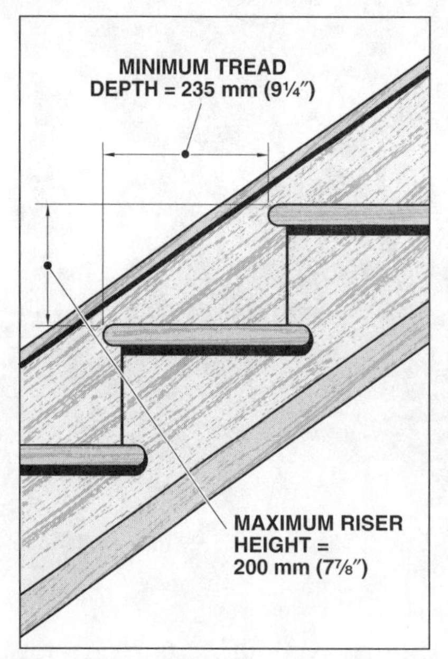

Figure 64-4. *The maximum riser height for residential stairways is 200 mm (7 7/8″) and the minimum tread depth is 235 mm (9 1/4″).*

Stairway Profiles. Stairway profiles include tread nosings and riser angles. The nosing of a tread is not permitted to have a radius larger than 25 mm (1″). Nosings should not be beveled more than 25 mm (1″). See **Figure 64-5.** For stairways with solid risers, a nosing should not extend from the riser less than 25 mm (1″).

Risers can be vertical or angled back from the underside of the leading edge of a tread. A nosing is not required if the riser height is angled back a minimum of 25 mm (1″).

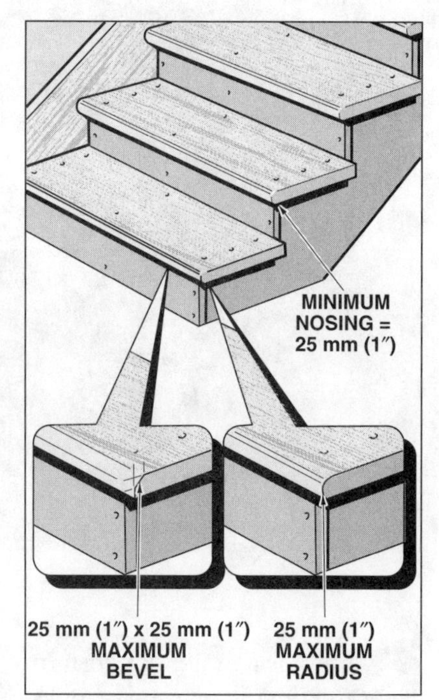

Figure 64-5. *Tread nosings are not permitted to have a bevel or radius greater than 25 mm (1″).*

The art of stairbuilding dates back to 6000 BC, when stairways were originally developed as exterior additions. In those days, stairways were constructed from notched logs.

Landings. A floor or landing must be constructed at the top and bottom of a stairway. However, a floor or landing is not required at the top of an interior stairway when a door does not swing over the stairway. Landings must be at least as wide as the stairways they serve and have a minimum dimension of 860 mm (34″) measured in the direction of travel for interior stairs and 900 mm (35″) for exterior stairs. See **Figure 64-6.** When constructing a straight-run stairway, the dimension of the landing in the direction of travel does not need to be greater than 1 100 mm (43″). A flight of stairs for a residential stairway should have a 3 700 mm (12′-2″) maximum total rise between landings or floors.

Intermediate stringers may be required depending on building code requirements.

The bottoms of balusters must be accurately cut at an angle for proper fit.

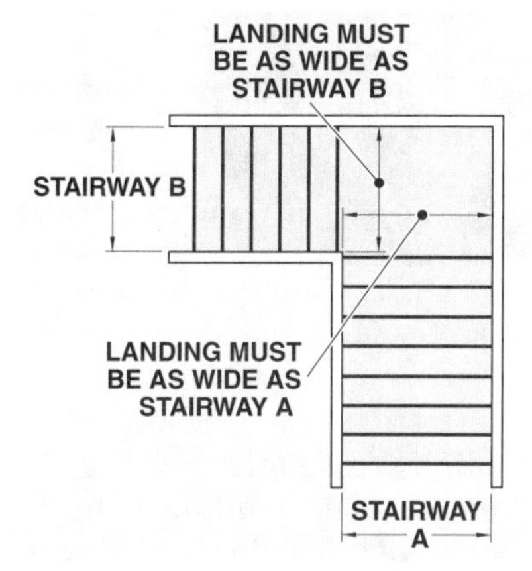

Figure 64-6. *Landings must be as wide as the stairway they serve and must be at least 860 mm (34″) wide.*

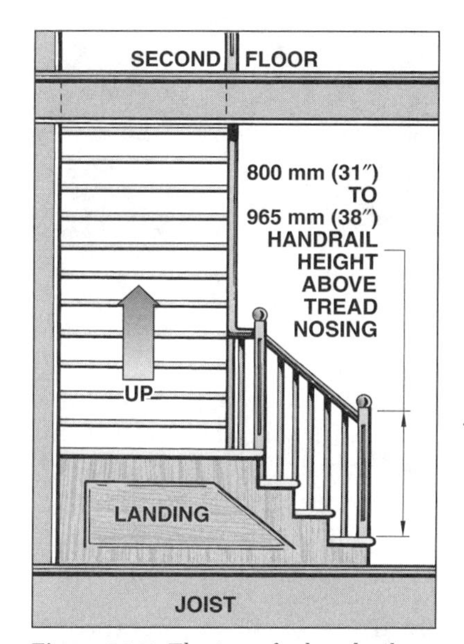

Figure 64-7. *The top of a handrail must be 800 mm (31″) to 965 mm (38″) above the tread nosings.*

Handrails. Handrails must be provided on at least one side of a stairway with four or more risers. The handrail height must not be less than 800 mm (31″) or more than 965 mm (38″), measured from the tread nosings. See **Figure 64-7.** Handrails for residential stairways should be continuous along the entire length of the stairway from a point directly above the top riser to a point directly above the lowest riser. Handrails may be interrupted by a newel post at a turn in the stairway. Handrail ends must be returned or terminate at a newel post or safety terminal. A clearance of at least 50 mm (2″) must be provided between the handrail and wall or other surface.

Handrails must be easily graspable. The width or diameter must be of a comfortable size, or the handrail must be shaped to provide a finger recess.

Commercial Stairway Requirements

The *National Building Code of Canada* addresses several stairway requirements including stairway width, headroom, treads and risers, stairway profiles, landings, total rise, circular and spiral stairways, and handrails.

Stairway Width. The stairway width for commercial buildings is based on building use, the number of occupants for which the building is designed, whether the stairway is equipped with a sprinkler system, and whether the stairway is to be accessible by individuals with disabilities. Buildings involved in the manufacturing, processing, or storage of hazardous materials require wider stairways than typical office buildings for faster egress (exit). The minimum stairway width for public buildings is 900 mm (35″). However, wider stairways may be required based on the number of occupants, the type of building occupancy, and whether the stair is a primary exit stair or an interior service stairway.

The National Building Code establishes minimum requirements to safeguard public health, safety, and welfare through structural strength, means of egress, and adequate light and ventilation. It also establishes minimum requirements to protect the safety of life and property against fire and other hazards.

Headroom Requirements. The minimum headroom for commercial stairways is 2 050 mm (6′-9″) measured vertically from a line connecting the edges of the nosings. The minimum headroom must be maintained the full width of the stairway and any associated landings.

Ramps may be installed to provide access for individuals with disabilities. Ramps should have a maximum 8.3% (1 in 12) slope.

Handrails must be easily graspable and shaped to provide a finger recess.

Riser Heights and Tread Depths. Riser heights for commercial stairways are 125 mm (4⅞″) minimum and 200 mm (7⅞″) maximum with the riser heights measured vertically between the leading edges of adjacent treads. In addition, the variation in riser heights between the largest and smallest riser height should not be more than 6 mm (¼″). Tread depths must be at least 250 mm (9⅞″) with the tread depths measured horizontally from the leading edges of adjacent treads. The variation in tread depth between the largest and smallest tread depth should not be more than 6 mm (¼″).

Stairway Profiles. The leading edge of a tread is not permitted to have a radius larger than 10 mm (⅜″) or less than 6 mm (¼″). Nosings should not be beveled more than 10 mm (⅜″) or less than 6 mm (¼″). Risers can be vertical or angled from the underside of the leading edge of a tread. If angled risers are specified, the angle should provide a back slope of not less than 25 mm (1″). Nosings must be of uniform size, including the leading edge of the floor at the top of the stairway. The underside of nosings should not be abrupt.

Landings. A floor or landing must be constructed at the top and bottom of a stairway. Landings must be at least as wide as the stairways they serve and have a minimum dimension measured in the direction of travel equal to the stairway width. When constructing a straight-run stairway, the dimension of the landing in the direction of travel does not need to be greater than 1 100 mm (43″). A flight of stairs for a commercial stairway should have a total rise not greater than 3 700 mm (12′-2″) between landings or floors.

Circular and Spiral Stairways. Stairs that are not part of a primary exit may incorporate curved or spiral sections, but primary exit stairs may not be spiral stairs. Curved exit stairs are acceptable, provided that

The curved riser is a prefabricated component that is constructed by sawing kerfs in the back and forming the riser around a template.

the minimum run at the inside radius is not less than 240 mm (9½″) and that the riser height and tread depth conform to straight stair requirements as measured 230 mm (9″) from the narrow end of the treads.

Handrails. Commercial stairways more than 1 100 mm (43″) wide are required to have handrails on each side. Handrail height must not be less than 865 mm (34″) or more than 965 mm (38″) measured from the tread nosings. Intermediate handrails are required on stairways over 2 200 mm (86″) wide so that no part of the stairway is more than 1 100 mm (43″) from a handrail. Handrails should be continuous on at least one side of the stairway, including landings, except where interrupted by a newel post or doorway. Where handrails terminate, they must extend at least 300 mm (12″) beyond the top riser and at least 300 mm (12″) plus the depth of one tread beyond the bottom riser.

For optimum graspability, cylindrical handrails must have a minimum outside diameter of 30 mm (1¼″) and a maximum outside diameter of 43 mm (1¾″). For noncylindrical handrails, the handrails must have a perimeter dimension greater than 100 mm (4″) and not more than 125 mm (5″) with a maximum cross-section dimension of 45 mm (1¾″). A clearance of at least 50 mm (2″) must be provided between the handrail and wall or other surface.

Stairwell Opening

A stairwell opening must be framed in the floor at the top of the stairway. In new construction, the width and length of a stairwell opening is shown in the prints. In remodeling work, a stairwell opening may have to be cut into an existing floor.

The width of a stairwell opening should be the same as the rough width of the stairs. The stairway opening length must be calculated to ensure the proper amount of headroom between the lower steps of the stairway and the end of the opening above. See **Figure 64-8.**

Straight-Flight Stairways

A straight-flight stairway is the simplest type of stairway to build. **Figure 64-9** shows a procedure for constructing a straight-flight stairway from the first to second story of a building.

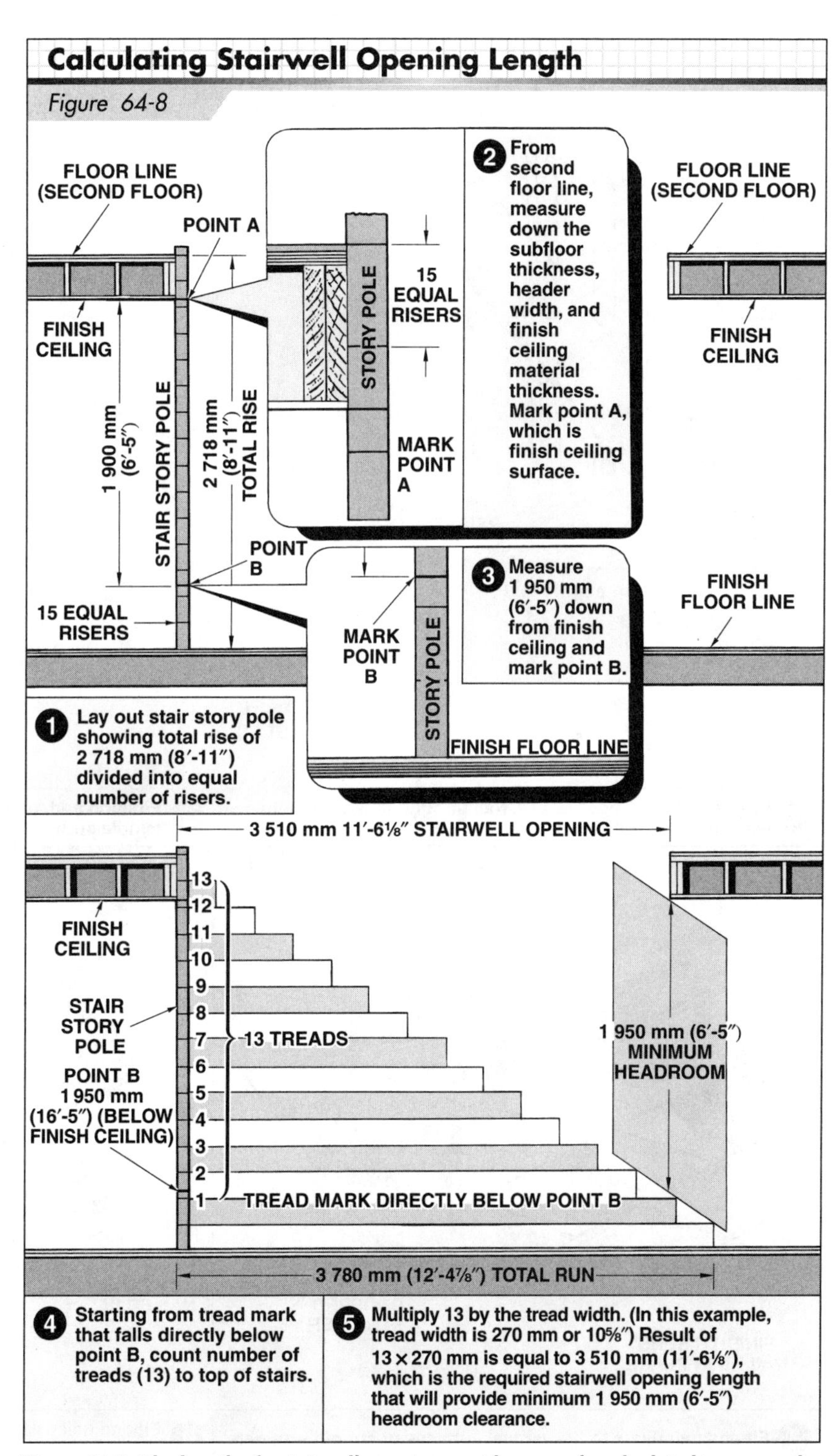

Figure 64-8. *The length of a stairwell opening must be properly calculated to ensure the proper amount of headroom.*

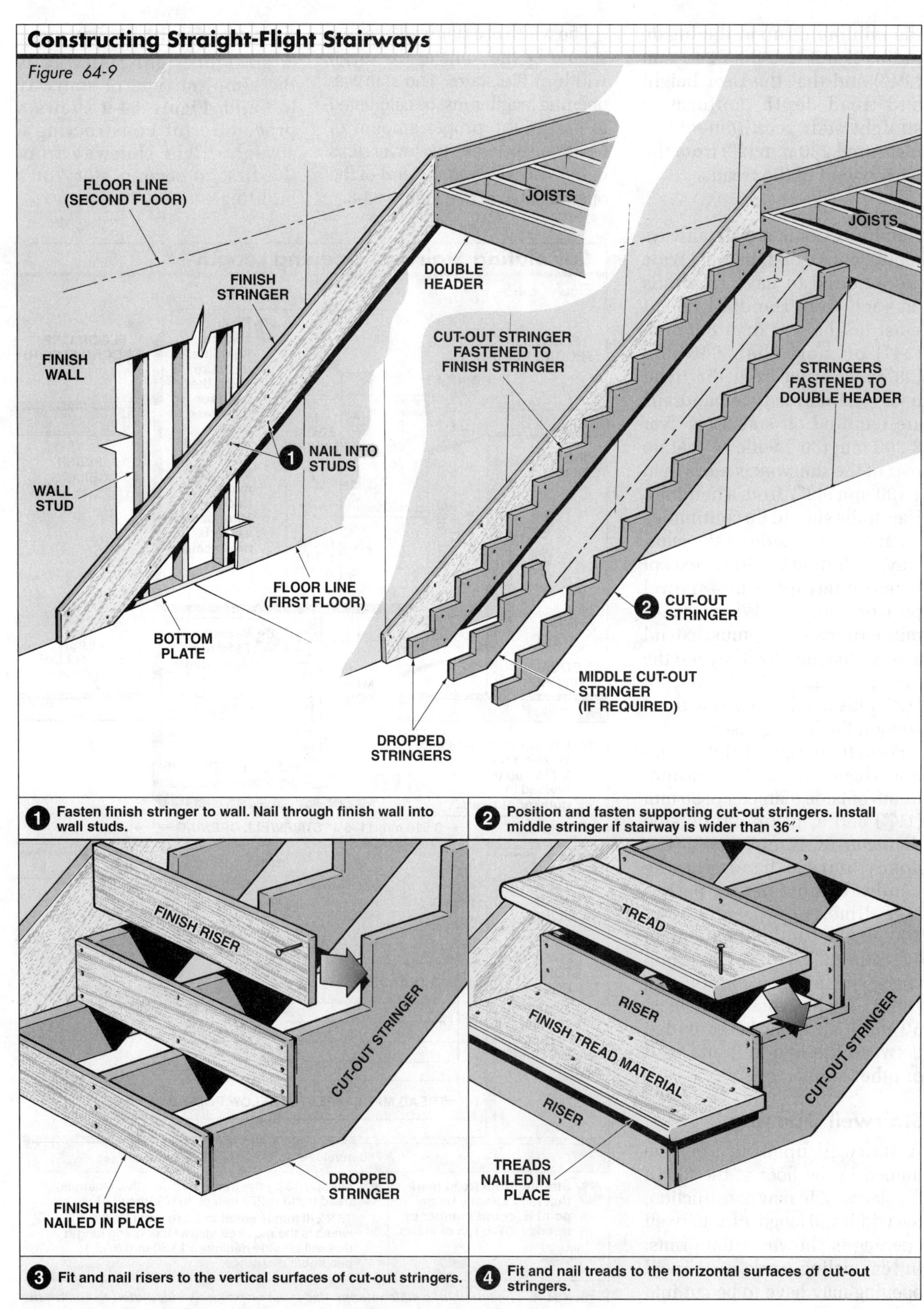

Figure 64-9. *A straight-flight stairway runs directly between different floor levels.*

For a straight-flight stairway with a landing, the first step is to frame the landing. The stairway is then constructed against the landing. **Figure 64-10** shows a procedure for constructing an L-shaped stairway with a landing. The stairway in this example is open on one side.

Stairways with Winders

Most stairways with winders are L-shaped. Instead of a landing separating the flights, a series of winder treads are used to make the turn. The winder section of most L-shaped stairways consists of two or three treads.

Stairways with winders are not as safe as straight-flight stairways. The depth of a winder tread varies from one end to the other. Stairways with winders are usually installed only where space does not permit a straight-flight stairway with a landing. See **Figure 64-11.**

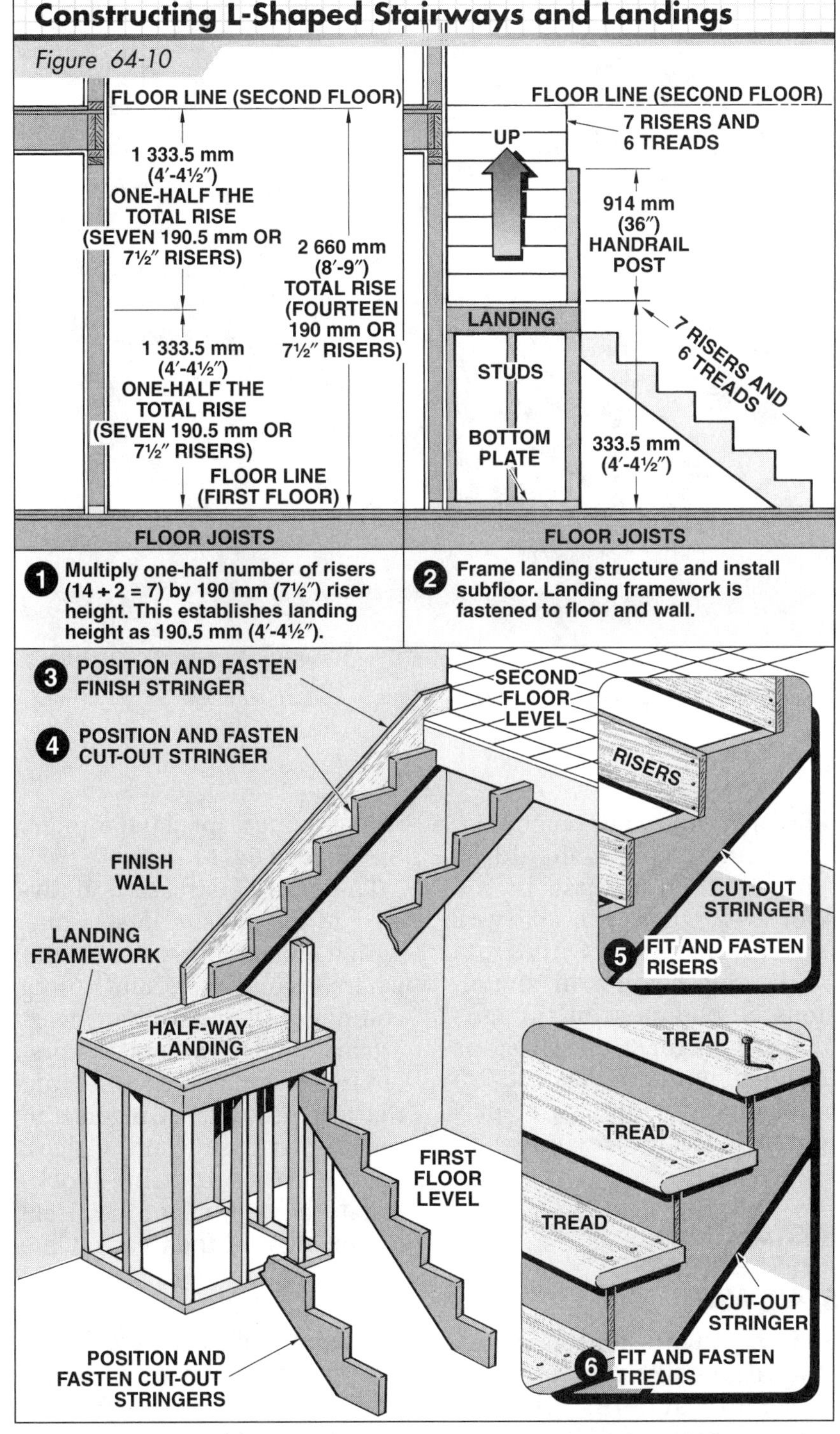

Figure 64-10. *When constructing an L-shaped residential stairway with a landing, the landing is built first. Stringers from the first floor to the landing, and from the landing to the second floor, are then installed.*

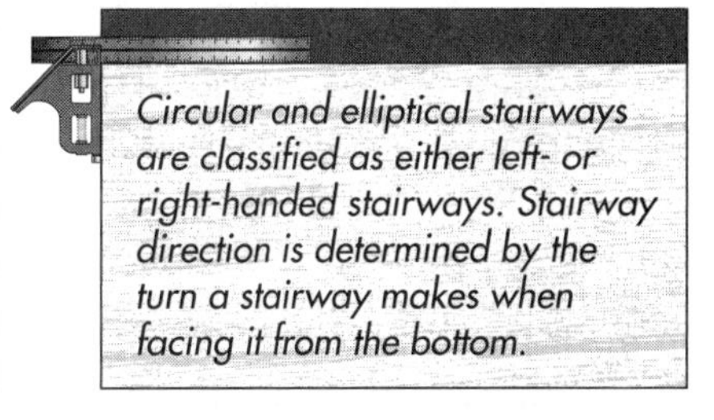

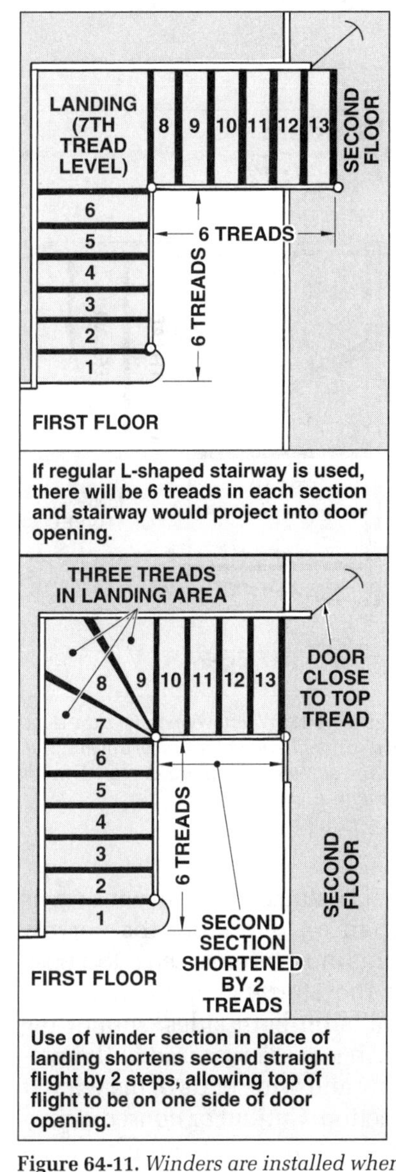

Figure 64-11. *Winders are installed when space does not allow a straight-flight stairway with a landing.*

An important dimension that must be established in designing a stairway with winders is the *line of travel*. See **Figure 64-12.** The line of travel is measured from the turn at the narrow ends of the winder steps, and is the place where a person is likely to walk when using the stairway. Winder treads must have a 235 mm (9¼″) minimum tread depth measured at a right angle to the tread's nosing at the line of travel. Individual winder treads are limited to a minimum angle of 30° and a maximum angle of 45°. A set of winder treads may not turn through more than 90° in total.

Semicircular landings are used to separate flights of stairs in this commercial stairway.

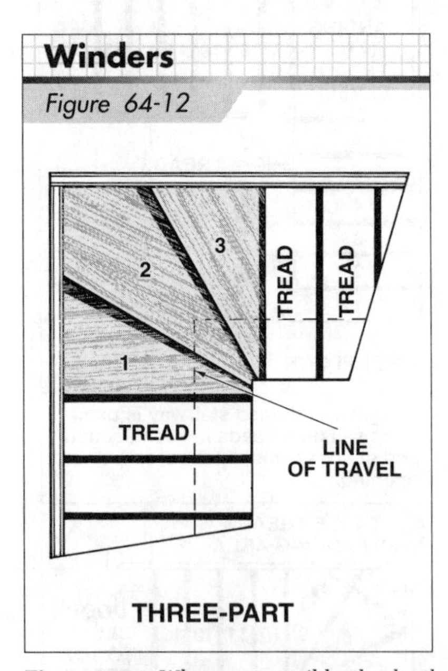

Figure 64-12. *Whenever possible, the depth of a winder tread at the line of travel should be the same as the tread depth along the straight-flight section.*

If a stairway with winders is open on one side, the narrow ends must be mortised into a post. If the stairway is enclosed by walls on both sides, supporting stringers are required at the narrow and wide ends of the winder section. Cut-out or housed stringers can be used. Prefabricated winder stairways are usually built with housed stringers.

Before a stairway is installed, the winder section should be laid out to full scale in the floor area where the stairway is to be installed. Only from a full-scale layout can dimensions be obtained for the cuts of the stringer. General steps in the layout and construction of a stairway with a three-part winder section are shown in **Figure 64-13.**

INSTALLING PREFABRICATED STAIRWAYS

Stairways with housed stringers are usually prefabricated in a shop and installed by carpenters who work for a stairway subcontractor. Grooves in the stringers are routed with a router using a special template. See **Figure 64-14.**

The entire staircase is delivered to the job site in sections, which include the stringers, finish treads and risers, and railing components. Stairway stringers are installed first. Next, the precut treads and risers are set into the grooves of the stringers and secured with glue and wedges. See **Figure 64-15.** Glue blocks fasten the bottoms of the risers to the edges of the treads. Glue blocks (instead of nails) are used with housed stringers to reduce the amount of squeaking. If one side of the stairway is open, a *mitered stringer* is used in which the corner joints of the risers and stringers are mitered. See **Figure 64-16.**

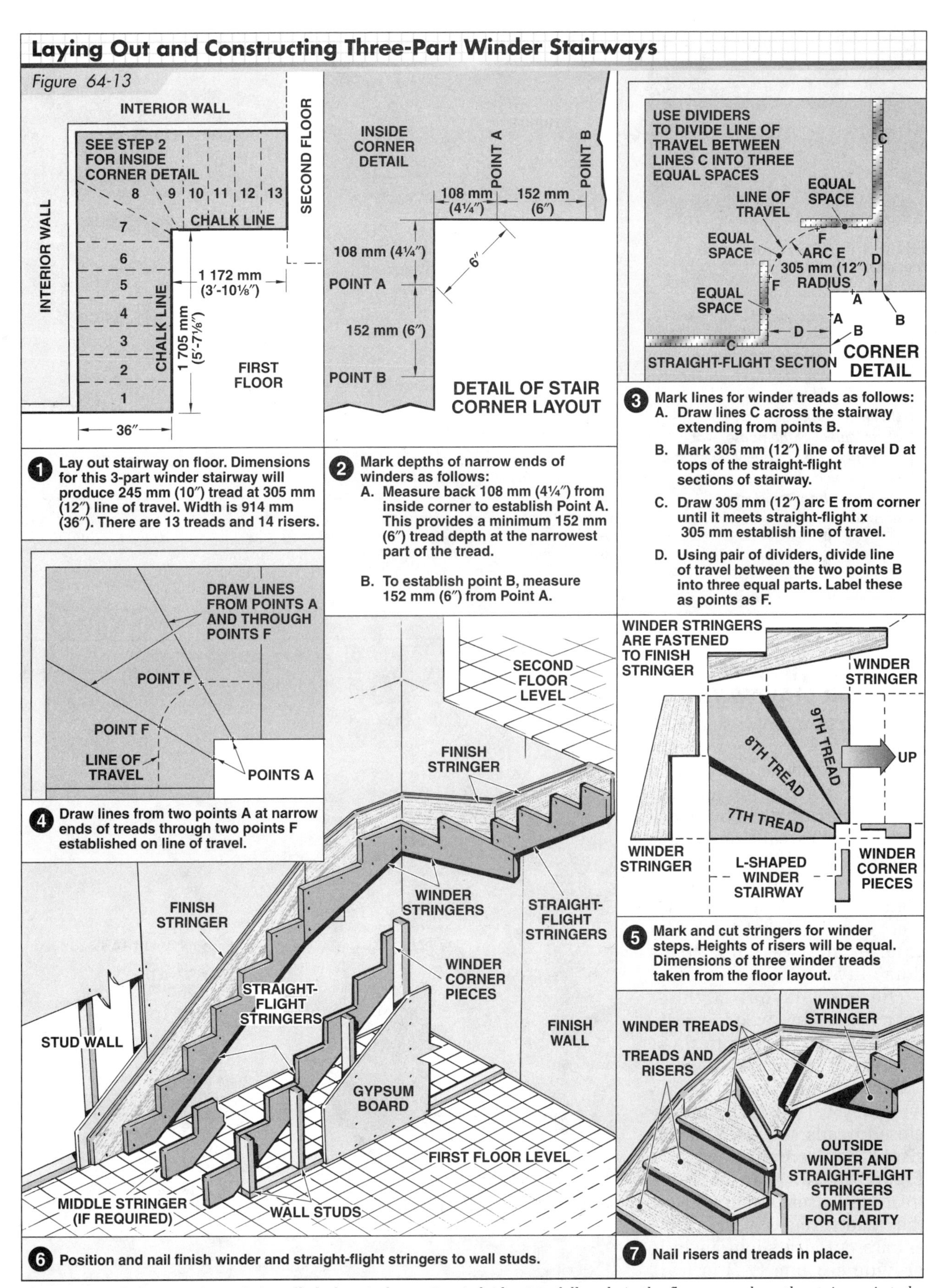

Figure 64-13. *Before a stairway is installed, the winder section is laid out to full scale in the floor area where the stairway is to be installed. In this example, the line of travel is 356 mm (14″). The tread depth along the winder and straight-flight line of travel will be 254 mm (10″). The total width of the stairway is 914 mm (36″).*

Figure 64-14. *Grooves in a housed stringer are made with a router.*

"Squeaky" treads and risers are a common complaint from occupants or owners of houses with older stairways. Squeaks may be a result of settlement of the house, wood shrinkage, and frequent use of the stairs, which causes separation of the treads from the stringers and/or risers.

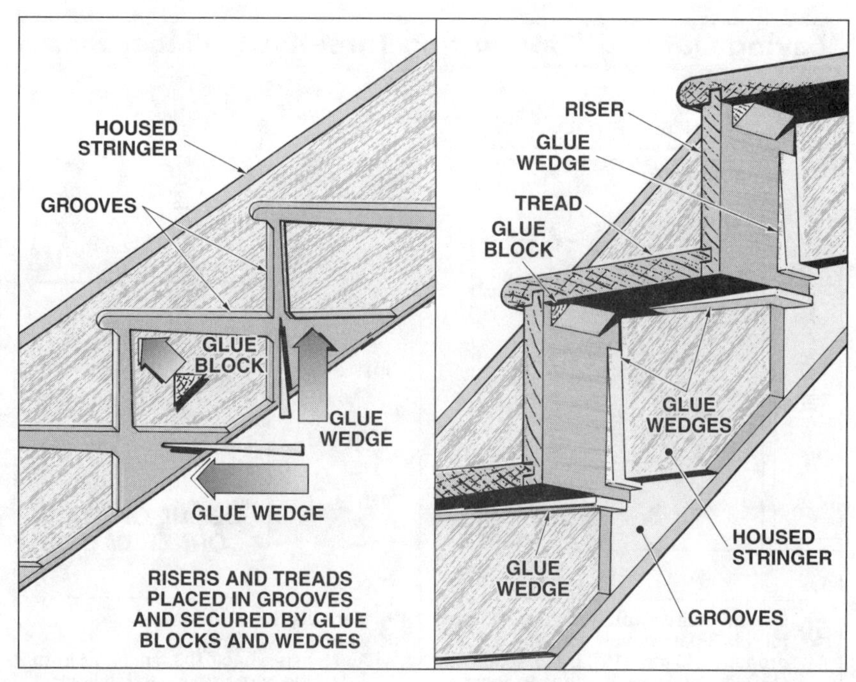

Figure 64-15. *When assembling a stairway with housed stringers, the risers and treads are placed in the grooves and secured with glue and wedges. Glue blocks fasten the bottoms of the risers to the edges of the treads.*

CONSTRUCTING EXTERIOR STAIRWAYS

A variety of exterior stairways are used for access to front and rear entrances, decks, and porches. An exterior stairway may be constructed entirely of wood, or its treads and risers may be finished with stone, tile, or other nonwood material. See **Figure 64-17.** Concrete is also a common material to use for exterior stairways due to its durability.

The basic layout methods for constructing wood exterior stairways are similar to those for interior stairways.

The steps of an exterior stairway may be finished with enclosed treads and risers, or the risers may remain open. See **Figure 64-18.** If an enclosed tread and riser design is used, the tread surfaces of the stringers are cut at a slight angle to provide a 3 mm (1/8") forward slope to provide proper water drainage.

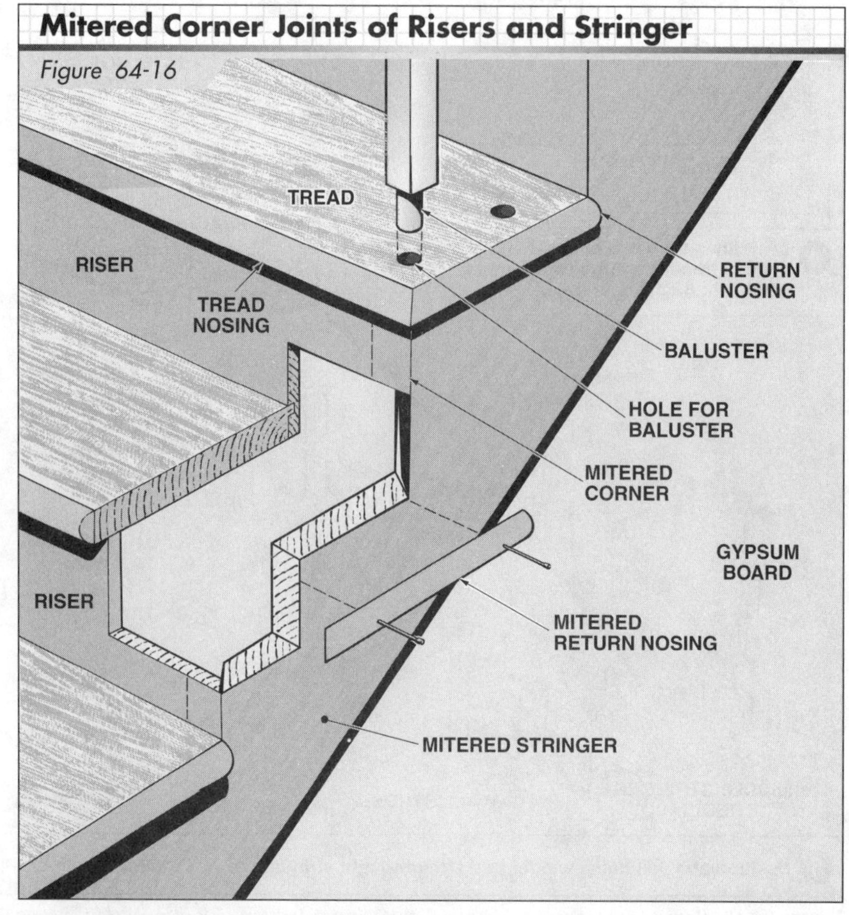

Figure 64-16. *In a mitered stringer, the corner joints of the risers and the stringer are mitered.*

A concrete bottom step is recommended for wood stairways if the bottom of the stringer does not rest on a concrete slab. If wood posts are used as part of the stairway structure, the posts should be supported by concrete piers that extend below the frost line.

One procedure for constructing an exterior stairway is shown in **Figure 64-19.** In this example, the stairway leads to a landing at the rear of the building.

Figure 64-17. *An exterior stairway may be constructed of wood or may be finished with nonwood materials.*

Exterior Tread and Riser Designs

Figure 64-18

TREAD
RISER
3 mm (1/8") FORWARD SLOPE FOR WATER DRAINAGE
WATER DRAINAGE
CUT-OUT STRINGER IS CUT WITH 3 mm (1/8") SLOPE
ENCLOSED TREAD AND RISER

SPLIT TREADS
OPEN RISERS
TREAD GAP
CUT-OUT STRINGER
WATER DRAINAGE
OPEN TREAD

Figure 64-18. *The steps of an exterior stairway may be finished with enclosed treads and risers or the risers may remain open.*

Ramps may be constructed adjacent to a building to provide access for individuals with disabilities. The National Building Code of Canada *outlines the requirements for ramps used as a means of egress. Items such as ramp slope, cross slope, vertical rise, minimum dimensions, landings, and changes in direction are addressed.*

Exterior stairways must be properly supported at the top.

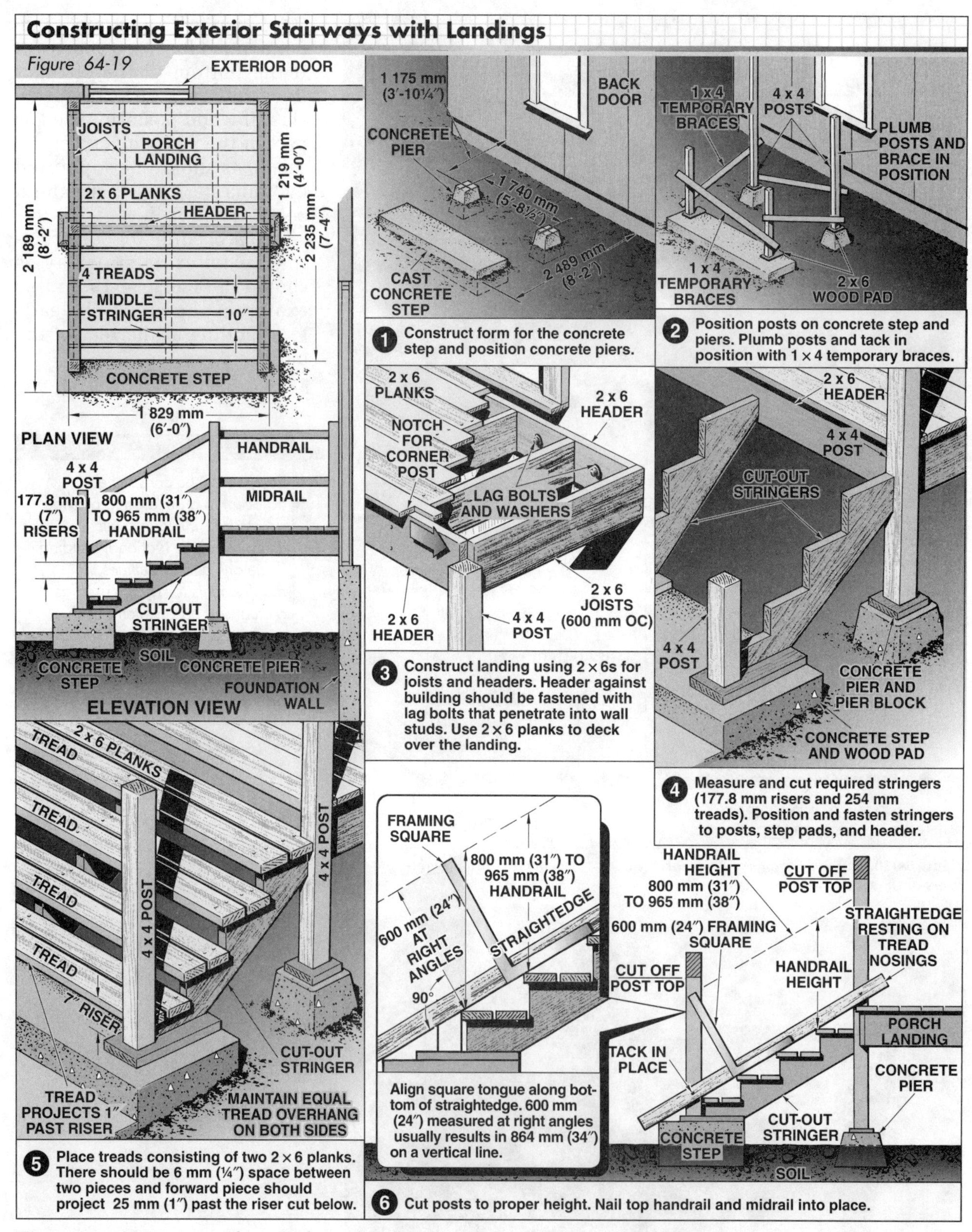

Figure 64-19. *Stairways must be properly supported. The stairway in this example is supported by a concrete step at the bottom, concrete piers, and a 2 × 6 header at the top. Local building codes may require the foot of the stairs and the posts to be supported on footings extending below the frost line.*

Quick Quiz®

Refer to the Interactive CD-ROM for the Quick Quiz® questions related to section content.

Post-and-Beam Construction

Page 713
Securing Ridge Ends of Transverse Beams

To meet Canadian codes, the sill plate shown in Figure 65-12 must be set in caulking, or on a gasket, to prevent air leakage. If the SIP is clad with OSB or plywood, it must be isolated from concrete within 200 mm (7⅞″) of ground.

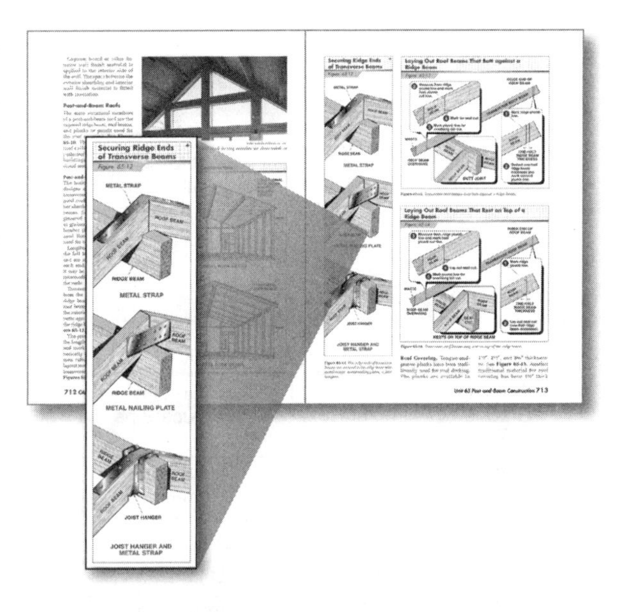

Building with Glulam Lumber

Page 725
Glulam Floor, Roof Beam, and Purlin Sizes

Figure 66-6 refers to imperial measurements and loading that may differ from Canadian standards. Some tables for glulam beams are found in the *National Building Code of Canada;* much more information is available from Canadian Wood Council publications such as *Wood Reference Handbook*.

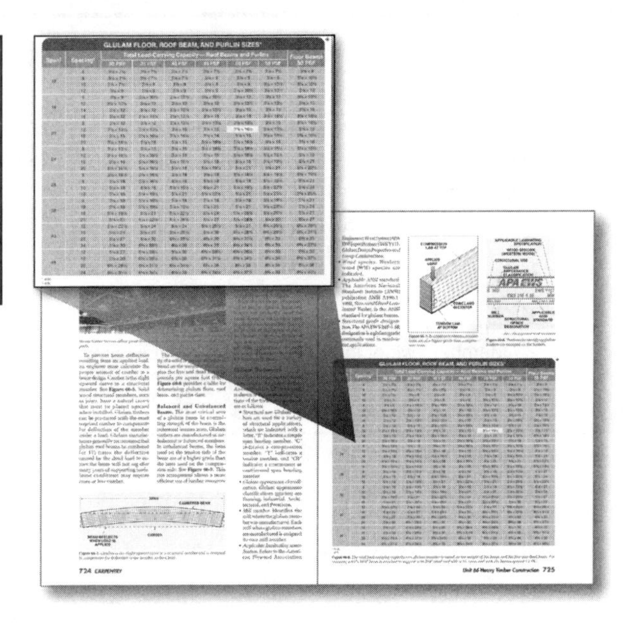

Commercial Species for Canadian Glulam

Commercial Species Group Designation	Species in Combination	Wood Characteristics
Douglas Fir-Larch (D.Fir-L)	Douglas fir, western larch	Woods similar in strength and weight. High degree of hardness and good resistance to decay. Good nail holding, gluing and painting qualities. Colour ranges from reddish-brown to yellowish-white.
Hem-Fir	Western hemlock, amabilis fir, Douglas fir	Lightwoods that work easily, take paint well and hold nails well. Good gluing characteristics. Colour range is yellow-brown to white.
Spruce-Pine	Spruce (all species except coast sitka spruce) lodgepole pine, jack pine	Woods of similar characteristics, they work easily, take paint easily and hold nails well. Generally white to pale yellow in colour.

Glulam Stress Grades

Stress Grade		Species	Description
Bending Grades	20f-E and 20f-Ex	D.Fir-L or Spruce-Pine	Used for members stressed principally in bending (beams) or in combined bending and axial load.
	24f-E and 24f-Ex	D.Fir-L or Hem-Fir	Specify EX when members are subject to positive and negative moments or when members are subject to combined bending and axial load such as arches and truss top chords.
Compression Grades	16c-E	D.Fir-L	Used for members stressed principally in axial compression, such as columns.
	12c-E	Spruce-Pine	
Tension Grades	18t-E	D.Fir-L	Used for members stressed principally in axial tension, such as bottom chords of trusses.
	14t-E	Spruce-Pine	

Standard Glulam Widths

Initial width of glulam stock		Finished width of glulam stock	
mm	in.	mm	in.
89	3-1/2	80	3
140	5-1/2	130	5
184	7-1/4	175	6-7/8
235 (or 89 + 140)	9-1/4 (or 3-1/2 + 5-1/2)	225 (or 215)	8-7/8 (or 8-1/2)
286 (or 89 + 184)	11-1/4 (or 3-1/2 + 7-1/4)	275 (or 265)	10-7/8 or 10-1/4
140 + 184	5-1/2 + 7-1/4	315	12-1/4
140 + 235	5-1/2 + 9-1/4	365	14-1/4

Notes:
1. Members wider than 365mm (14-1/4") are available in 50mm (2") increments but require a special order.
2. Members wider than 175mm (6-7/8") may consist of two boards laid side by side with longitudinal joints staggered in adjacent laminations.

Glulam Appearance Grades

Grade	Description
Industrial Grade	Intended for use where appearance is not primary concern such as in industrial buildings; laminating stock may contain natural characteristics allowed for specified stress grade; sides planed to specified dimensions but occasional misses and rough spots allowed; may have broken knots, knot holes, torn grain, checks, wane and other irregularities on surface.
Commercial Grade	Intended for painted or flat-gloss varnished surfaces; laminating stock may contain natural characteristics allowed for specified stress grade; sides planed to specified dimensions and all squeezed-out glue removed from surface; knot holes, loose knots, voids, wane or pitch pockets are not replaced by wood inserts or filler on exposed surface.
Quality Grade	Intended for high-gloss transparent or polished surfaces, displays natural beauty of wood for best aesthetic appeal; laminating stock may contain natural characteristics allowed for specified stress grade; sides planed to specified dimensions and all squeezed-out glue removed from surface; may have tight knots, firm heart stain and medium sap stain on sides; slightly broken or split knots, slivers, torn grain or checks on surface filled; loose knots, knot holes, wane and pitch pockets removed and replaced with non-shrinking filler or with wood inserts matching wood grain and colour; face laminations free of natural characteristics requiring replacement; faces and sides sanded smooth.

Excerpt from *Wood Reference Handbook*, ©1991 by Canadian Wood Council

Formwork Construction

Page 754
Standard Metric Rebar Sizes

Metric rebar designation indicates the diameter of the rebar in millimetres, rounded to the nearest 5 mm.

METRIC REBAR SIZES

Size	Mass*	Diameter†	Cross-Section Area‡
#10M	0.785	11.3	100
#15M	1.570	16.0	200
#20M	2.355	19.5	300
#25M	3.925	25.2	500
#30M	5.495	29.9	700
#35M	7.850	35.7	1000
#45M	11.775	43.7	1500
#55M	19.625	56.4	2500

* in kg/m
† in mm
‡ in mm^2

Prestressed Concrete

Page 773
Canadian Precast/Prestressed Concrete Institute (CPCI)

The Canadian Precast/Prestressed Concrete Institute is a trade association representing manufacturers of concrete products. More information is available at **www.cpci.ca.**

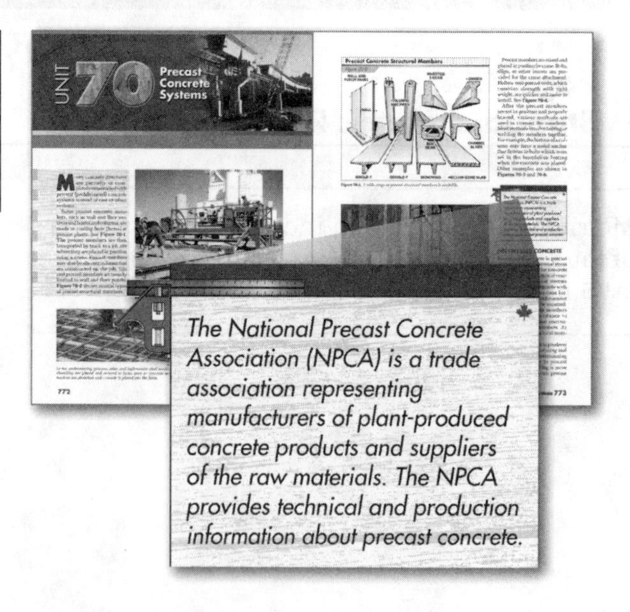

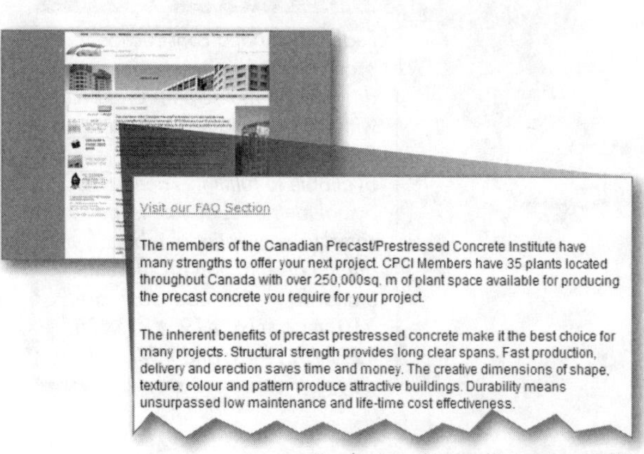

Canadian Precast/Prestressed Concrete Institute (CPCI)

See Carpentry CD-ROM Canadian Resources–Canadian Precast/Prestressed Concrete Institute (CPCI)

Page 785
Lumber/Timber Grades

The Canadian Wood Council publishes information on Canadian dimension lumber grades and uses in *Wood Reference Handbook.*

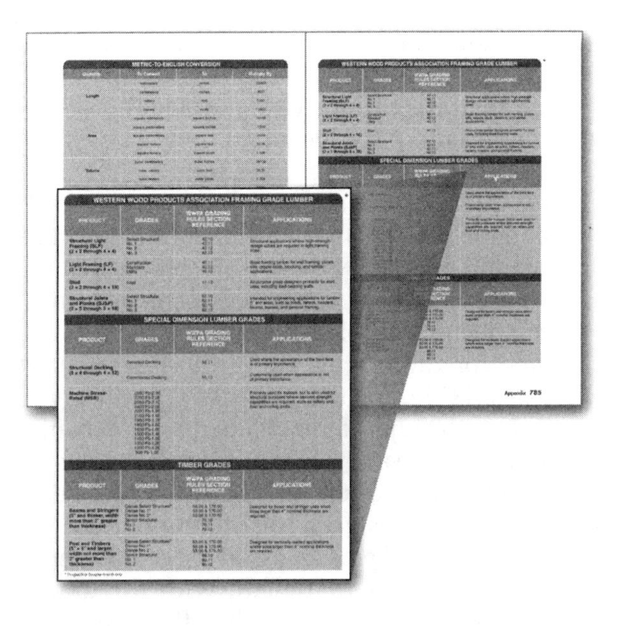

Specialty Products

Machine Stress-Rated Lumber (MSR Lumber)

Lumber which is evaluated mechanically is called machine stress-rated (MSR) lumber. Canadian MSR lumber is manufactured in conformance with the National Lumber Grades Authority (NLGA) *Special Product Standard 2 (SPS-2).*

Unlike visually graded lumber where the anticipated strength properties are determined from assessing a piece on the basis of appearance, the strength characteristics of MSR lumber are determined by applying forces to a member and actually measuring the stiffness of a particular piece. Because the stiffness of each piece is measured, and because the strength is measured on some pieces through a quality control program, MSR lumber can be assigned higher design stresses than visually graded dimension lumber.

In addition to the information shown on a grade stamp for visually graded lumber, the MSR grade stamp indicates the grade by reference to bending strength (Fb) and stiffness (E).

As lumber is fed continuously into the mechanical evaluating equipment, stiffness is measured and recorded by a small computer, and strength is assessed by correlation methods. MSR grading can be accomplished at speeds up to 365 m (1000′) per minute including the affixing of an MSR grade mark.

MSR lumber is also visually checked for properties other than stiffness that might affect the suitability of a given piece.

The mechanical analysis of MSR lumber results in the affixing of an f-E stamp to a piece of lumber. The f-E classifications relate directly to allowable values. For example, an 1800f – 1.6E grade has been tested for an allowable stress in bending of 12.4 MPa (1800 psi) and a modulus of elasticity (E) of 11000 MPa (1.6 million psi) under normal duration of load.

MSR lumber is favoured particularly by specialized users such as truss manufacturers where higher strength per volume of lumber and reliability resulting from measured design values is required.

Fingerjoined Lumber

Fingerjoined lumber is dimension lumber into which finger profiles have been machined. The pieces are then end glued together.

With fingerjoining, the length of a piece of lumber is not limited by tree size. In fact, the process may result in the production of joists and rafters in lengths of 12 m (40′) or more.

Although fingerjoining is used in several wood product manufacturing processes including the horizontal joists for glulam manufacture, the term fingerjoined lumber applies to dimension lumber.

Canadian fingerjoined lumber is manufactured in conformance with National Lumber Grades Authority (NLGA) *Special Product Standard 1 (SPS-1)* or *Special Product Standard 3 (SPS-3).* Fingerjoined lumber produced to the requirements of SPS-1 is interchangeable with non-fingerjoined lumber of the same grade and strength. For example, a No.2

fingerjoined joist may be assigned the same engineering properties as a No.2 joist made from one continuous length.

The fingerjoining process allows the removal of strength-reducing defects to produce a product with higher engineering properties. The strength of the joints is controlled by stipulating the quality of wood that must be present in the area of the joint.

For example, Select Structural, No.1, and No.2 grade joints must be formed in sound, solid, pitch-free wood that meets the slope of grain and other general requirements of the grade.

For all other grades, joints are formed in wood that meets visual requirements for No.2 or Standard grades, and the surface area of the joint having pitch or honeycomb must not exceed 10% of the area.

Each piece must be comprised of species from the same species group, and strict tolerances are established for the machining of the fingers, the quality, the mixing, and the curing of the adhesive. Depending on the type of fingerjoined lumber being manufactured, edge and flat bending tests and tension tests are performed on each piece to ensure the joint can meet the design value of the lumber.

Fingerjoined lumber is assessed for visual grade and for machine-tested strength, with the grade assigned being the lower of either the visual or the stress grade. As mentioned previously, there are two standards for Canadian fingerjoined lumber. One covers bending members (SPS-1) used in either the horizontal or vertical position, and the other covers members intended for use in a vertical position in stud walls (SPS-3).

All fingerjoined lumber manufactured to the Canadian NLGA standards carries a grade stamp indicating:

- the species or species combination information
- the seasoning designation (S-Dry or S-Green)
- the registered symbol of the grading agency
- the grade
- the mill identification
- the NLGA standard number and the designation CERT FGR JNT (certified finger joint) or Cert Fin Jnt Vertical Use Only (certified finger joint for vertical use only).

Timber

Grading of Timbers

Both categories of timbers, Beams and Stringers, and Posts and Timbers, contain three stress grades: Select Structural, No.1, and No.2, and two non-stress grades (Standard and Utility).

The stress grades are assigned design values for use as structural members. Non-stress grades have not been assigned design values.

No.1 or No.2 are the most common grades specified for structural purposes. No.1 may contain varying amounts of Select Structural, depending on the manufacturer. Unlike Canadian dimension lumber, there is a difference between design values for No.1 and No.2 grades for timbers.

Select Structural is specified when the highest quality appearance and strength are desired.

The Standard and Utility grades have not been assigned design values. Timbers of these grades are permitted for use in specific applications of building codes where high strength is not important, as in the case of deck uses, blocking, or short bracing.

Timbers are generally not grade marked, since they are surfaced rough and may be used in exposed locations. If needed, a mill certificate may be obtained to certify the grade.

Cross cutting can affect the grade of timber in the Beams and Stringers category because the allowable size of knot varies along the length of the piece (a larger knot is allowed near the ends than in the middle). Therefore, for this grade category, the timber must be regarded if it is cross cut.

Sizes Available

Timbers are sometimes manufactured to large dimensions and resawn later to fill specific orders.

Moisture Content

The large size of timbers makes kiln drying impractical due to the drying stresses that would result from differential moisture contents between the interior and exterior of the timber. For this reason, timbers are usually dressed green (moisture content above 19%), and the moisture content of timber upon delivery will depend on the amount of air drying which has taken place.

Like dimension lumber, timber begins to shrink when its moisture content falls below about 28%. The degree of shrinkage depends on the climatic conditions of the environment. For example, timbers exposed to the outdoors usually shrink from 1.8% to 2.6% in width and thickness, depending on the species. Timbers used indoors, where the air is often drier, experience greater shrinkage, in the range of 2.4% to 3.0% in width and thickness. Length change in either case is negligible.

When constructing with Posts and Timbers or Beams and Stringers, allowance should be made for anticipated shrinkage based on the moisture content at the time of assembly. Where the building envelope relies on caulked seals between timbers and other building components, the selection of caulks should take into account the amount of movement that must be accommodated as shrinkage occurs.

Minor checks on the surface of a timber are common in most service conditions and, therefore, an allowance has been made for them in the assignment of working stresses. Checks in columns are not of structural importance unless a check develops into a rough split that will divide the column.

Excerpt from *Wood Reference Handbook,* pp. 125-128. ©1991 by Canadian Wood Council.

Canadian Dimension Lumber—Grades and Uses

Grade Category	Grades	Common Grade Mix	Principal Uses
Structural Light-Framing 38 to 89mm (2" to 4" nom.) thick 38 to 89mm (2" to 4" nom.) wide	Select Structural No.1 No.2 No.3	No.2 and Better	Used for engineering applications such as for trusses, lintels, rafters and joists in the smaller dimensions.
Structural Joists and Planks 38 to 89mm (2" to 4" nom.) thick 114mm (5" nom.) or wider	Select Structural No.1 No.2 No.3	No.2 and Better	Used for engineering applications such as for trusses, lintels, rafters, and joists in the dimensions greater than 114mm (5" nom.).
Light Framing 38 to 89mm (2" to 4" nom.) thick 38 to 89mm (2" to 4" nom.) wide	Construction Standard Utility	Standard and Better (Std. & Btr.)	Used for general framing where high strength values are not required such as for plates, sills, and blocking.
Stud 38 to 89mm (2" to 4" nom.) thick 38mm (2" nom.) or wider 3m (10') or less in length	Stud Economy Stud		Made principally for use in walls. Stud grade is suitable for bearing wall applications. Economy grade is suitable for temporary applications.

Notes:

1. Grades may be bundled individually or they may be individually stamped but they must be grouped together with the engineering properties dictated by the lowest strength grade in the bundle.
2. The common grade mix shown is the most economical blending of strength for most applications where appearance is not a factor and average strength is acceptable.
3. Except for economy grade, all grades are stress graded which means specified strengths have been assigned and span tables calculated. Economy and utility grades are suited for temporary construction or for applications where strength and appearance are not important.
4. Construction, Standard, Stud, and No.3 grades should be used in designs that are composed of 3 or more essentially parallel members (load sharing) spaced at 610mm (2') centres or less.
5. Strength properties and appearance are best in the premium grades such as Select Structural.

Excerpt from *Wood Reference Handbook,* ©1991 by Canadian Wood Council

Page 786
Appearance Lumber Grades

The Canadian Wood Council publishes information on appearance and grades of softwoods used for interior finish in *Wood Reference Handbook.*

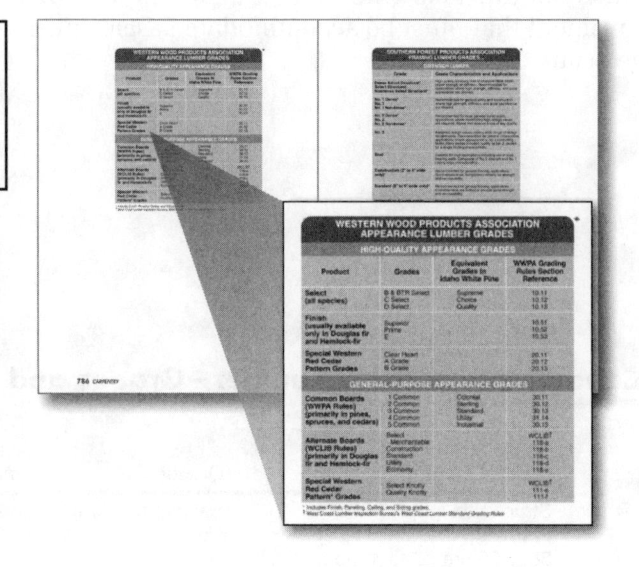

WESTERN WOOD PRODUCTS ASSOCIATION
APPEARANCE LUMBER GRADES

Product	Grades	Equivalent Grades in Idaho White Pine	WWPA Grading Rules Section Reference
HIGH-QUALITY APPEARANCE GRADES			
Select (all species)	B & BTR Select C Select D Select	Supreme Choice Quality	10.11 10.12 10.13
Finish (usually available only in Douglas fir and Hemlock-fir	Superior Prime E		10.51 10.52 10.53
Special Western Red Cedar Pattern Grades	Clear Heart A Grade B Grade		20.11 20.12 20.13
GENERAL-PURPOSE APPEARANCE GRADES			
Common Boards (WWPA Rules) (primarily in pines, spruces, and cedars)	1 Common 2 Common 3 Common 4 Common 5 Common	Colonial Sterling Standard Utility Industrial	30.11 30.12 30.13 30.14 30.15
Alternate Boards (WCLIB Rules) (primarily in Douglas fir and Hemlock-fir	Select Merchantable Construction Standard Utility Economy		WCLIB† 118-a 118-b 118-c 118-d 118-e
Special Western Red Cedar Pattern* Grades	Select Knotty Quality Knotty		WCLIB† 111-e 111-f

* Includes Finish, Paneling, Ceiling, and Siding grades.
† West Coast Lumber Inspection Bureau's *West Coast Lumber Standard Grading Rules*

Page 787
Framing Lumber Grades

The Canadian Wood Council publishes information on appearance and grades of softwoods used for interior finish in *Wood Reference Handbook.*

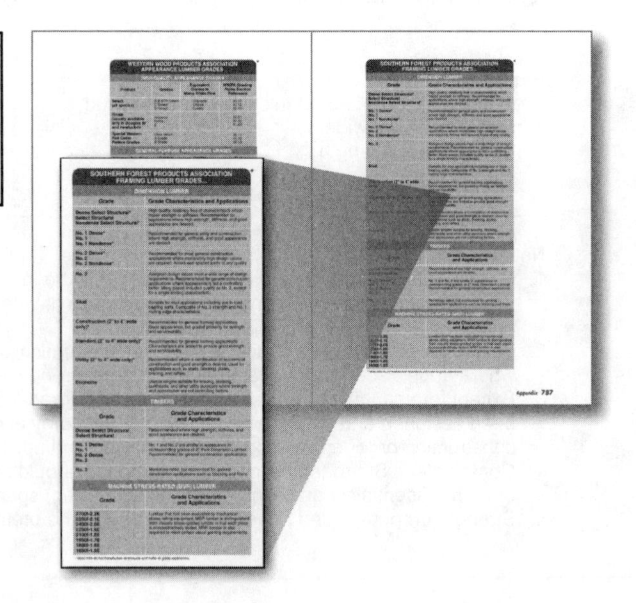

SOUTHERN FOREST PRODUCTS ASSOCIATION
FRAMING LUMBER GRADES

Appearance Characteristics of Softwoods used for Interior Finish

Species	Characteristics
Western red cedar	Reddish brown, close even grain, annual layers easily distinguishable, relatively soft.
Douglas fir	Heartwood bright red to yellow, sapwood nearly white, relatively free of resins, pronounced variable rings, hard surface.
Pacific coast hemlock	Heartwood reddish brown, sapwood nearly white, close straight grained, relatively free of resins, hard surface.
Lodgepole pine	Heartwood light brown to red, sapwood nearly white, close straight grained, relatively soft.
Sitka spruce	Heartwood light reddish brown, sapwood nearly white, medium-coarse grain, fairly durable.
Eastern white pine	Heartwood cream white, sapwood mainly white, ring width variable close straight grain, fairly resinous, relatively soft.
Jack pine	Heartwood bright red, sapwood yellow to white ring width variable pronounced saps, fairly resinous, fairly durable.
Red pine	Heartwood bright red, sapwood yellow to white ring width variable pronounced saps, fairly resinous, fairly durable.
Eastern white cedar	Heartwood light reddish brown, sapwood nearly white, close straight grained, relatively soft.

Suggested Softwood Grades for Interior Finish Work

Species	Grade Category	Grades	Characteristics
All species except Western red cedar	K.D. (kiln dry) Finish	C and Better	Small imperfections, suitable for clear finishing.
		D	Larger imperfections, suitable for painted surfaces.
	K.D. Ceiling & Siding	C and Better D	Same as for K.D. Finish. Same as for K.D. Finish.
	K.D. Casing & Base	C and Better D	Same as for K.D. Finish. Same as for K.D. Finish.
	Industrial Clears	B and Better	Supreme grade, clear on 1 face.
		C	Small imperfections.
		D	Utility grade for surfaces to be painted.
Eastern white pine and red pine	Selects	B and Better	Many pieces are completely clear.
		C	Small imperfections, suitable for clear finishing.
		D	More imperfections, suited for less exacting applications.
Western red cedar	Finish, Panelling, and Drop Siding K.D.	Clear Heart	Many pieces are completely clear.
		A	Small imperfections, suited for clear finishes.
		B	More imperfections, short lengths defect free.
	Industrial Clears	B and Better	Supreme grade, clear on 1 face.
		C	Small imperfections.
		D	Utility grade for surfaces to be painted .
	Tight Knotted Panelling and Siding	Select	Best quality where knotty appearances desired.
		Utility	Some loose knots permitted.

Page 790
APA Trademarks

The Canadian Wood Council publishes information on Canadian lumber grade stamps in *Wood Reference Handbook.*

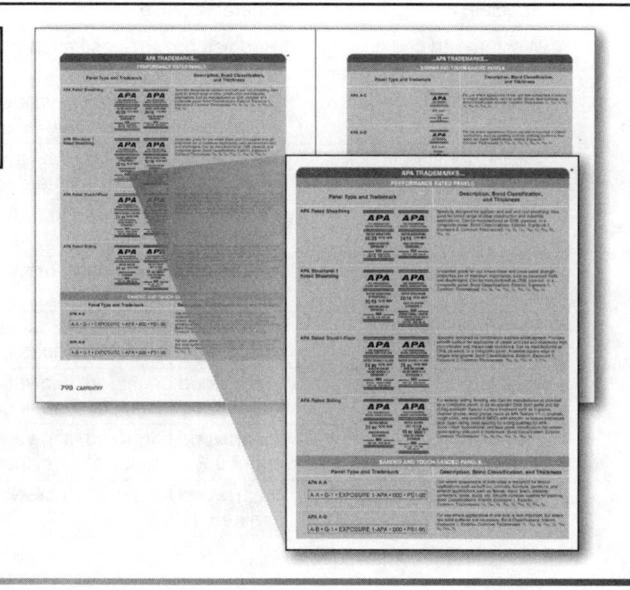

Canadian Lumber Grade Stamps

A.F.P.A.® 00 — Mill designation (Grading agency)
S—P—F — Species group
S-DRY — Moisture content
No. 1 — Assigned Grade

Alberta Forest Products Association
11738 Kingsway Avenue, Suite 200
Edmonton, AB T5G 0X5
780-452-2841
www.abforestprod.org
(Approved to supervise fingerjoining and machine stress-rated lumber)

A.F.P.A.® 00
S—P—F
S-DRY
No. 1

Association des manufacturiers de bois de sciage du Québec (Québec Lumber Manufacturers' Association)
5055 boul. Hamel ouest, bureau 200
Québec, QC G2E 2G6
418-872-5610
www.sciage-lumber.qc.ca
(Approved to supervise fingerjoining and machine stress-rated lumber)

Canadian Lumbermen's Association
27 Goulburn Avenue
Ottawa, ON K1N 8C7
613-233-6205
www.cla-ca.ca
(Approved to supervise fingerjoining and machine stress-rated lumber)

Cariboo Lumber Manufacturers' Association
197 Second Avenue North, Suite 205
Williams Lake, BC V2G 1Z5
250-392-7778
www.clma.com
(Approved to supervise fingerjoining and machine stress-rated lumber)

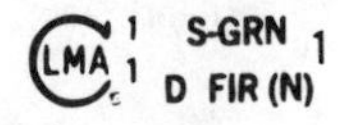